Marine treatment of sewage and sludge

Marine treatment of sewage and sludge

Proceedings of the conference organized by
the Institution of Civil Engineers and held
in Brighton on 29-30 April 1987

Thomas Telford, London

Organizing committee: D. O. Lloyd (Chairman), I. D. Brown, Dr J. A. Charlton, Dr T. J. Lack, F. N. Midmer, N. Paul, J. M. Reynolds, J. K. Waterhouse, J. A. Young

British Library Cataloguing in Publication data
Marine treatment of sewage and sludge
1. Sewage sludge. Marine treatment
I. Title
628.3

ISBN 0 7277 1301 9

First published 1988

Published for the Institution of Civil Engineers by Thomas Telford Ltd, Telford House, 1 Heron Quay, London E14 9XF.

Printed and bound in Great Britain by Robert Hartnoll (1985) Ltd, Bodmin, Cornwall.

Contents

ECONOMIC CONSIDERATIONS

THE WAY AHEAD

Keynote address

THE RT HON. LORD NATHAN, Chairman, Sub-Committee G, Select
Committee on European Communities, House of Lords

This opening address concerns itself firstly with how laws
are made. Pressures on parliamentary time are not
sufficiently understood. They often preclude logic and
create delays. Environmental regulation is about risk, and
public perception of risk relates closely to catastrophe and
crisis, and parliament reflects that perception. So
government is far more likely to act to prevent the
repetition of an event which is unlikely to recur rather than
one which certainly will.

There is increasing concern about vehicle emissions and
their effect on man and the environment. Regulations have
been made and there is a lot of discussion about the damage
they can cause; but there is not a single case in which
anybody has died as a consequence of what comes out of the
back end of a car. Hundreds of thousands are killed by the
front end; and yet nothing is done about it.

The Clean Air Act is another example of how catastrophe and
crisis can produce swift action.

For some years there was an appalling sulphurous smog that
descended upon London in the winter. The air could scarcely
be breathed, one's throat was sore and one's eyes streamed
and visibility was reduced to about 5 yards. Every time this
happened - and it was practically every winter - over 40 000
elderly people died.

When in 1952 the great national cattle show was held in
London and the cows all died as a result of the smog, it was
the pressure of that catastrophe, and not the continuous
experience of 40 000 human deaths per year, that gave rise to
the Clean Air Act of 1956. This has produced a remarkable
transformation for London and other cities.

It is this need to heighten awareness by creating a feeling
of crisis that results in pressure groups. That essentially
is their task.

Concern was expressed at one time that the concentration of
lead in air was resulting in adverse effects on the central
nervous system of small children and on their intelligence.
As a result the Lawther Committee, made up of mostly medical
experts, investigated this. They considered that there was
no certain and cogent evidence justifying the concern which

Marine treatment of sewage and sludge. Thomas Telford Ltd, London, 1988

had been expressed. Than a brilliantly led pressure group
went into action to bring about the elimination of lead from
petrol.

At this time the Royal Commission on Environmental
Pollution, of which I was then (and am now) a member, was
thinking about preparing a report which would include this
subject but would also be more wide-ranging. In the light of
the concern created by the pressure group we decided urgently
to carry out and publish a study on lead in the environment —
and of course lead in petrol was a key element in our
enquiry. We agreed with the conclusion reached by the
Lawther Committee — that there was no cogent evidence of
damage to young children from lead in the air. What we did
find, however, was that the margin of safety between the
levels of lead in the air and the levels at which damage
might be caused were much too narrow, and far narrower than
in the case of other toxic substances. We also found that
the burden of lead in the environment was increasing at a
faster rate than was wise and that this rate of increase
should be reduced as far as practicable. The easiest and
cheapest way to achieve both these objectives would be to
eliminate lead from petrol. We therefore made that
recommendation. Our report was published at 2.30 pm one
afternoon and a government statement was made in Parliament
accepting that recommendation at 3.30 pm the same day.
Further, the government, in accordance with our
recommendations, immediately put in hand an initiative with
the Commission in Brussels, since this was a matter for
European as well as UK legislation.

The Royal Commission on Environmental Pollution would not
have acted with the speed they did — nor would government
have reacted so quickly — had there been no action by the
pressure group.

The European Community is not affected by parliamentary
time pressures. Generally, proposals of the Commission do
not derive from heavenly inspiration reduced to draft
legislation by the pure light of reason. They derive often
enough from a strongly expressed point of view by a member
state or indeed a pressure group — and there is nothing wrong
in that.

An example of how community legislation comes into
existence can be illustrated as follows. The Greens in
Germany are far more extensive in number and influence than
the formal Green Party. They have great political influence.
Accordingly a course of action may be proposed by them and
may be accepted as policy by the German Government. The
German Government are then confronted by German industry,
whose costs, it is said, will be substantially increased by
adoption of the environmental controls proposed which are
therefore unacceptable unless accepted throughout the
Community. The German Government therefore instigates
Community action through the Commission in Brussels and the
Commission produces draft legislation giving effect to the

proposal which originated with the Greens. Of course, this
is a very simplified version of what has happened on a number
of occasions in the past, but it gives some idea as to what
actually happens and why certain parts of the proposed
Community legislation come forward.

An example is to be found in the Community proposal
relating to dumping of waste at sea resulting, so far as
sewage is concerned, in a reduction in dumping of 10% per
annum for five years. The idea lying behind the whole
proposal is the adoption of the precautionary approach (the
Vorsorgeprinzip) which is a concept favoured not only in
Germany but also elsewhere, such as in the Netherlands.

At its most extreme it is said that the sea is such a
vulnerable environment that it should not be used at all for
waste disposal. It is said that scientific knowledge about
the effects of waste disposal to sea is incomplete;
monitoring of waste disposal sites may show no evidence of
significant pollution but this does not prove that the
practice is safe. It merely shows that waste is diluted and
dispersed on a wide scale and adds to the general stress on
the marine environment. If evidence comes to light which
shows that waste has a damaging effect it will be too late.

So the question arises of whether we should accept the idea
lying behind the proposal even if we have to fight perhaps
for a much longer transitional period to allow for the
radical changes which would be required here, or whether we
should not accept the principle at all. An alternative view
is that one can start with the proposal that material that
man puts into the sea does not constitute pollution unless it
has some damaging effect.

Consideration then has to be given to the assimilative
capacity of the receiving environment. For example, many
wastes currently discharged to sea are dispersed and diluted
to low concentrations so that they have no detectable effect
on the environment except sometimes around the point of
discharge. According to this approach the limits of safe
discharge are governed by the assimilative capacity of the
receiving environment.

However agreeable it may be to contemplate life without
waste - or sewage - this is not going to happen. It will
have to be disposed of, whatever advances in technology in
reducing wastes and recycling may be achieved. So the
question is what is the best thing to do with it or, to use
the phrase originated by the Royal Commission on
Environmental Pollution, what is the best practicable
environmental option? To discard dumping sludge at sea from
the available options is absurd in the absence of any cogent
evidence of environmental objections (let alone expense) of
disposal to land or incineration.

I have used the phrase 'cogent evidence' (in relation to
environmental damage) because I believe, in a proper context,
the precautionary principle has an important part to play in
pollution control. This was referred to in the tenth report

3

of the Royal Commission on Environmental Pollution, where it was said that the state of knowledge of the risks of damage varies widely, yet it is not always or indeed often possible to delay any response until there is scientific certainty. The politician or manager who must decide what action to take immediately cannot wait for the rigorous proof that is properly demanded by a referee for a scientific journal. For those responsible for the protection of the public and the environment there is a difference between what can be believed with confidence and what in the absence of certainty it is prudent to assume. The manager must determine whether or not the perceived risk of damage justifies the cost of its alleviation; if it does, then a policy for the abatement of the pollution must be instituted.

I found myself subsequently faced with this problem in relation to air pollution and acid rain on which my Committee was preparing a report. I thought that action should be taken notwithstanding the absence of scientific proof and set out criteria which I thought important and which led me to that conclusion. It seemed to me that the magnitude of the damage resulting, the length of time over which it might become apparent, and its widespread effects – if the fears expressed proved to be justified – impelled one to act now.

The justifiable costs of taking action against the perceived danger and the resultant damage should be measured.

This approach was of course far removed from that of the CEGB, until recently, in maintaining that no action might be taken on emissions from coal-fired power stations in the absence of scientific proof of damage caused by them.

I believe that the approach I have indicated may help to reconcile the contrasting approaches to pollution control to which I have referred.

1. European Community activities towards the protection of the marine environment

V. MANDL, Head of Division, Protection and Management of Water, Commission of the European Communities, Brussels

SYNOPSIS. The paper outlines the measures which have been taken by the European Community to control and reduce pollution of marine waters. These include directives on bathing water quality, shellfish waters, pollution by dangerous substances, exchange of information and a proposal on dumping at sea. Participation in discussions and initiatives with other international organisations is considered an important part of protection of the marine environment.

BACKGROUND

The need for a Community environmental policy was announced by Heads of States as early as 1972. Since then the interest of Member States in the protection of the environment has grown and from 1973 to 1986 three action programmes have been adopted. A fourth action programme covering the period 1987 - 1992 has been already presented and is presently being discussed. Over these years the measures taken by the Commission were developed in such a way as to prevent, as far as possible, pollution at source.

The measures to reduce marine pollution have centered on six priority areas:

1. The definition of quality objectives

Most Community seas are threatened by degradable pollutants which come from land-based scattered sources. In order to contain this pollution the Commission has proposed - and the Council adopted - a number of directives establishing quality objectives for water as a function of use.

Directives have been approved setting quality standards for some surface water[1], bathing water[2], water supporting fish[3], and shellfish culture[4].

2. The protection from dangerous substances

Other pollutants, which are more harmful because of their toxicity, persistence and bioaccumulation require more specific measures. It was this which brought the Council to adopt the framework directive for the protection of the aquatic environment against pollution by dangerous substances[5]. This has been followed by other 'daughter' directives for protection of the aquatic environment from specific dangerous substances.

3. The Protection of the Sea against Oil and Chemical Pollution

The "Torrey Canyon", "Amoco Cadiz" accidents and the blow-out of a platform in the North Sea are a few examples in a series of major incidents which prompted the Council to ask the Commission to draw up a Community action plan to combat marine oil pollution. The setting up of an EEC Advisory Committee on the control and reduction of oil pollution at sea (ACPH) and the establishment of a Community Information System for hydrocarbons, were two major events towards the prevention and combating of oil pollution.

Some elements of the Information System, and above all the inventory, are intended for use in emergency situations, while others, mainly information regarding impact, contain data of a rather general character and may be used in any circumstances. Unfortunately, oil pollution is not confined to sensational major spills like the "Amoco Cadiz" and the sea very often is regarded as an open sewer for dumping waste oils from ships, for fuel tank cleaning operations and in some cases for off-shore activities, etc. The Community being aware of these problems, put every effort into combating them with action at a wider international level and is signatory to a number of International Conventions dealing with the protection of the sea against pollution.

A proposal sent in August 1985 to the Council on dumping and incineration of waste at sea replaces an old proposal of 1976 and has as its main objective the prevention and reduction of marine pollution caused by the deliberate dumping of waste from ships and aircraft and also by the incineration of wastes at sea.

This new proposal takes into account a number of international Conventions, but envisages even tighter standards for the Community.

For example, it would prohibit the dumping at sea of certain wastes containing particularly harmful substances and the prohibition of the incineration of waste at sea should also be considered possible in the long term. In addition, there would be a system of prior authorisation for all dumping and incineration of wastes and materials, obviously except those which are banned. This authorisation should reduce the total quantities of certain substances dumped by 50% in five years starting from 1 January 1990.

Radioactive substances are not covered. They may be dealt with in separate proposals at a later date but this must depend on developments elsewhere - in particular under the London Dumping Convention. The Commission has already proposed that the Community participate alongside Member States in that Convention.

4. Data collection and exchange of information

A Council decision this year was to extend the Community Information System to cover other harmful substances as well as hydrocarbons. The information necessary for the control and reduction of pollution caused by a significant discharge at sea of hydrocarbons is already available to the relevant authorities of the Member States. Information concerning other harmful substances is not expected to be available before late 1987.

In view of the importance of sewage treatment and the degree of treatment given to sewage by Member States the Commission has funded a study to establish and update statistical data on sewage treatment in the European Community. The object being the collection of statistical data on sewage treatment in public facilities compiled by the national administrations responsible for sewage purification in the Member States and to present this data in a coherent and global form.

A point of interest in the statistical data on sewage treatment was a global balance of the gross and net organic load discharged into the environment.

The percentage of the population served by sewage networks in the ten Member States is given in Fig. 1. Fig. 2 gives a treatment capacity per inhabitant and population. Fig. 3 shows the municipal sewage treatment plants and summarizes the treatment capacity and the annual capital expenditure on the treatment plants of each Member State (excluding Spain and Portugal). All of these data are for public treatment plants.

The Commission acts as a central focus of information for other data also. Data is fed into the Commission on, for example, bathing water, and on programmes of reduction of dangerous substances. The Commission makes information available in the form of reports and arranges seminars, workshops etc. on implementation of directives. In addition national research programmes on specific topics such as sewage sludge are coordinated by groups sponsored by the Commission which make regular publications of discussions held and reports of seminars.

5. Measures specific to certain branches of industry

A Directive related to waste from the titanium dioxide industry (78/176/EEC)[6] has been adopted with the ultimate aim of the elimination of pollution from this course. Member States have drawn up programmes for the progressive reduction of wastes from existing

Country	Year of latest available data	(1) Population served in % of total population	(2) Total population in 1980 (10³ inhabitants)	(3) = (1)x(2) Population served (10³ inhabitants)
B	1981	55 %	9,859	5,422
D	1980	88,6 %	61,566	54,547
DK	1977	92 %	5,123	4,713
F	-	-	53,714	(26,857)*
GR	-	-	9,599	(4,799)*
I	1971	56 %	57,070	31,959
IRL	1983	67 %	3,401	2,279
L	1983	96 %	365	350
NL	1980	90 %	14,150	12,735
UK	1980	95,9 %	56,010	53,714
EUR10	-	73 %	270,857	(197,375)

(...) = estimate
* assumption : p.s. is 50 % of total population.

Fig. 1. Population served by the sewerage network – summary of latest available data (8)

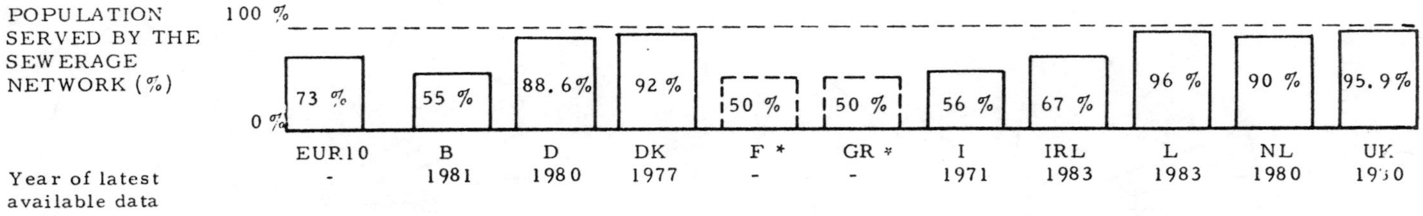

* Estimated

Fig. 2. Treatment capacity per inhabitant, and population served by the sewerage network - summary of latest available data (8)

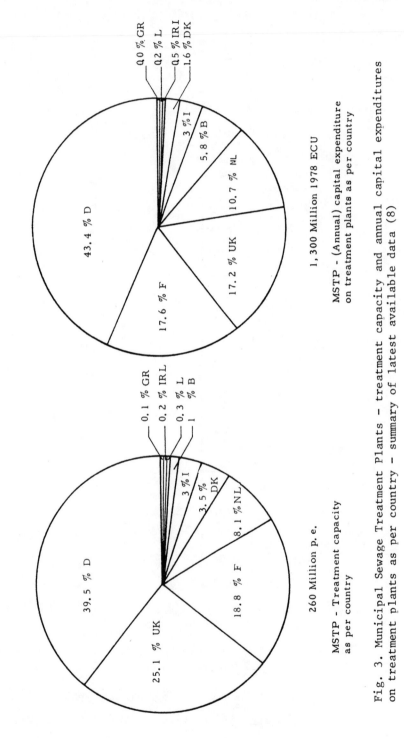

1,300 Million 1978 ECU

MSTP - (Annual) capital expenditure
on treatment plants as per country

260 Million p. e.

MSTP - Treatment capacity
as per country

Fig. 3. Municipal Sewage Treatment Plants – treatment capacity and annual capital expenditures
on treatment plants as per country – summary of latest available data (8)

establishments. These programmes have been forwarded to
the Commission which, after having examined them, has
proposed their harmonization in order to improve the
protection of the environment and to diminish the
distortion of competition.   This proposal has not yet
been adopted by the Council.   A follow-up Directive on
monitoring of the environment affected by wastes from
the titanium dioxide industry (82/883/EEC)[7] establishes
monitoring requirements and procedures for surveillance.

6. Activities under international agreements and
    organisations

The Community has become involved in, among other
things, the Bonn Agreement, the Paris Convention and the
Oslo Convention.
The Community has been directly involved in the work to
combat pollution of the Mediterranean sea since 1977
when it became a Contracting Party to the Barcelona
Convention and to its protocols:
- for the prevention of pollution by dumping from ships
  and aircraft;
- for protection against pollution from land-based
  sources;
- and, concerning specially protected areas.
Here, the Commission, apart from its role of co-
ordination between Member States, attaches great
importance to the proper implementation of the measures
taken under the Mediterranean Action Plan.   In
particular, it is actively involved in the realisation
of the priority objectives retained by the Contracting
Parties at their last meeting which took place in Genoa
in September last year. These objectives include:
- the establishment of reception facilities for ballast
  waters and other oily residues in ports of the
  Mediterranean;
- the establishment as a matter of priority of sewage
  treatment plants in all cities with more than 100,000
  inhabitants   and   appropriate   outfalls   and/or
  appropriate treatment plants for all towns with more
  than 10,000 inhabitants;
- compulsory environmental impact assessments;
- lessening the risk of maritime navigation accidents;
- the protection of the endangered marine species;
- concrete measures to achieve substantial reduction in
  pollution and solid waste discharges;
- identification and protection of marine and coastal
  historic sites and reserves;
- measures to prevent and combat forest fires, soil loss
  and desertification, and
- a reduction in air pollution
A special programme for Mediterranean Strategic Action
Plan (MEDSAP) was started in 1986 by the European
Commission to promote better environmental management
and protection in the Mediterranean area.   Ten pilot

projects concerning the collection of waste, the protection of endangered species and the establishment of oily waste reception facilities were launched in 1986.

OTHER MATTERS

There are further relevant aspects of Community environment policy which are important:

The Fourth Environment Action Programme sets the priorities for Community environment policy over the period 1987-1991. So far as the protection of the sea is concerned, the priorities will be the implementation of existing commitments and engagements rather than any major new initiatives. As in many other areas of Community environment policy, it is believed that it is now time to concentrate more on putting into effect what we already have on the stocks rather than further stretching the slender resources of the Commission staff in creating new legislation. In particular, attention will be concentrating on:

- the implementation of a strategic action plan for the Mediterranean;
- harmonized enforcement at Community level of optional annexes to the MARPOL Convention;
- the implementation of the Community Information System dealing, in particular, with harmful substances spilled at sea, and
- continuation of demonstration pilot projects for marine protection against oil and other chemical substances.

The issues involved in the disposal to sea of dredgings removed from coastal area are both fundamental and technical. The Commission has initiated a study into some aspects of contamination of sediments to better establish the case, or absence of a case, for Community measures on the matter.

European Year of the Environment (EYE), 1987, is a new initiative which is of course not limited to the protection of the sea but in which certainly this could play a part. The main aim of EYE is to promote public awareness and to persuade the public by means of specific activities and projects that protection of the environment is the concern of everyone.

CONCLUSION

Concern is growing in many Member States about pollution of the sea — the ultimate sink of most polluting emissions, whether they are initially discharged to water, to air or to land. It is evident that major reductions in marine pollution will be required to avoid the risk of overtaxing the sea's absorptive capacity.

Experts in the field of marine pollution are faced with a major technological challenge. More resources, both human and material, must be used to prevent pollution, to improve the present situation and to mitigate the effects of damage to the environment.

In particular, we must ensure that our contribution as a Community extends beyond the confines of the EEC and that we use our best endeavours to support and stimulate wider interest in the work of other international organisations. Pollution does not recognise frontiers or territorial limits. In declaring war upon it we need to ensure that we extend our alliances so that our response to the challenges we face is both powerful, flexible and comprehensive.

REFERENCES

[1] Council Directive of 16 June 1975 concerning the quality required of surface water intended for the abstraction of drinking water in Member States (OJ No. L 194, 25.7.1975, p. 26).

[2] Council Directive of 8 December 1975 concerning the quality of bathing water (OJ No. L 31, p. 1)

[3] Council Directive of 18 July 1978 on the quality of fresh waters needing protection or improvement in order to support fish life (OJ No. L 222, 14.8.1878, p. 1).

[4] Council Directive of 30 October 1979 on the quality required of shellfish waters (OJ No. L 281, 10.11.1979, p. 47)

[5] Council Directive of 4 May 1976 on pollution caused by certain dangerous substances discharged into the aquatic environment of the Community (OJ No. L 129, 18.5.1976, p. 23)

[6] Council Directive of 20 February 1978 on waste from the titanium dioxide industry (OJ No. L 54, 25.2.1978, p. 19)

[7] Council Directive of 3 December 1982 on procedures for the surveillance and monitoring of environments concerned by waste from the titanium dioxide industry (OJ No. L 378, 31.12.1982, p. 1)

[8] Commission of the European Communities "Updating of Statistical Data about Sewage Treatment in the European Community" Report May 1984

2. Legislation and future developments in the marine disposal of sewage and sludge in the USA

T. T. DAVIES, PhD, Director, Office of Marine
and Estuarine Protection, US Environmental Protection Agency

SYNOPSIS

This paper describes the history, status, and prognosis for marine disposal of sewage and sludge in the United States from a regulatory and environmental protection perspective. This brief look at marine disposal of sewage and sludge focuses geographically on the near coastal waters (NCW) of the United States. The NCW include the Great Lakes (Lakes Superior, Huron, Michigan, Erie, and Ontario), estuaries, and shallow coastal waters out to the 12 mile, contiguous zone limit.

INTRODUCTION

1. Regulation and management of Marine disposal of treated and untreated sewage and sewage sludge in the United States, indeed, water pollution control and water quality management is an evolutionary process. Through the early decades of water pollution control, up until 1970, we devoted our attention in the United States to identification and abatement of major water quality problems. Consistant with statutory authorities of these early years, regulatory controls on sources of pollution were based on theoretical relationships between pollutant discharges to surface water bodies and changes in water quality. Federal Government regulatory authorities during this time period were limited.

2. Then, in the early 1970's, we increased our commitment and shifted our emphasis, through the use of improved technologies and legal authorities, to intensive regulation of municipal and industrial (M & I) point sources, including the marine disposal of treated sewage and sludge. Consistent with changed statutory requirements and authorities, technology based minimum treatment requirements have been established for most M & I discharges to navigable waters

of the U.S. This increased commitment and change in water pollution control emphasis was accompanied by a significant increase in the responsibilities of several Federal agencies and the Environmental Protection Agency (EPA) in particular. Point source pollution control progress in the United States has been substantial over the past 15 years.

3. Now, we are assessing the status of water quality management in the United States; renewing our commitment towards achieving national water quality goals and objectives; and making hard decisions on where our future emphasis needs to be if we are to achieve these goals and objectives cost-effectively and within a reasonable time-frame. We are at a crossroad in our pollution control program and need to decide whether more stringent controls for point sources are needed to achieve water quality objectives in certain geographic areas or a broadened commitment of balanced controls for both point and nonpoint pollutant sources would be more cost effective.

HISTORICAL PERSPECTIVE... THE EARLY YEARS

4. Prior to 1970, marine disposal of sewage and sludge was regulated under the 1899 Rivers and Harbors Act, more commonly known as the 1899 Refuse Act; early versions of Federal water pollution control legislation; and limited authorities of the State and local governments. The 1899 Refuse Act outlawed discharges and deposits of industrial wastes into navigable waters of the United States without a permit, but organization of a Refuse Act Permit Program (RAPP) was not initiated until 1970. Even then, establishment of enforceable permit conditions was problematic, and the RAPP was not fully implemented.

5. The first Federal Water Pollution Control Act (FWPCA) was legislated and signed in 1948 and amended four times by 1966. Those early laws focused water pollution control efforts on identification and characterization of water quality problems for major river basins and water bodies of the United States, also, establishment of enforceable water quality standards, and preparation of pollution abatement plans. Most such pollution abatement plans were based on the assumption that compliance with water quality standards could be achieved by limiting discharges of relevant pollutants to calculated levels and that the calculated discharge limits could be achieved by imposing controls on M & I discharges. Establishment and enforcement of such water quality-based

pollutant discharge limits was complicated and slow because it was necessary to establish quantitative relationships between pollutant discharges and documented water quality problems that would withstand legal challenges in court. In other words, it was necessary, under the FWPCA, to demonstrate that water quality was being adversely affected by specified pollutant discharges before enforceable controls could be imposed on such discharges, including marine discharges of sewage and sludge.

6. Water pollution control technologies in those early years were limited. M & I treatment requirement tended to be narrowly focused on the so-called "conventional pollutants"... biochemical oxygen demand (BOD), suspended solids, pH, fecal coliforms, and oil and grease. More often than not, in fact, BOD was the primary focus of water quality-based pollutant control requirements. Calculations of pollutant discharge limits to meet water quality standards were usually restricted to calculations of allowable BOD discharges to meet dissolved oxygen requirements for the receiving waters. The calculated BOD limits were then allocated among identified M & I dischargers, resulting in calculated treatment requirements for each discharge. Such water quality-based treatment requirements were seldom calculated for nonconventional pollutants, such as nitrogen and phosphorus, and very seldom calculated for toxic pollutants.

7. In addition to water quality-based regulatory requirements, relatively small funding subsidies were provided under the FWPCA for the construction of publicly owned (that is, municipal) treatment works (POTW). However, the effectiveness of these funds with respect to pollution control requirements was not closely directed or regulated.

8. In summary, water pollution control efforts during the early years before 1970 emphasized water quality-based treatment requirements for M & I discharges in major river basins and focused on the control of conventional pollutants, BOD in particular. This was an important beginning and resulted in the construction of numerous industrial waste treatement plants and POTWs.

9. However, in 1970, the newly organized Council of Environmental Quality (CEQ), in its first annual report to Congress, documented widespread deterioration of water quality in the United States caused by the disposal of inadequately treated, oftentimes untreated industrial wastes, sewage and sludge. The CEQ recom-

mended legislative action to establish national water quality and pollution control objectives and regulatory authorities. The 1972 Clean Water Act (CWA) and the 1972 Marine Protection, Research and Sanctuaries Act (MPRSA) were legislated by Congress in response to the CEQ recommendations. The CWA regulates discharges of pollutants to United States surface waters, and the MPRSA regulates dumping of materials into open ocean waters. These laws have had a dramatic impact on regulation of pollutant sources including the marine disposal of sewage and sludge.

PROGRAM STATUS..."AT THE CROSSROADS"

10. The CWA established two national goals, attainment of "fishable and swimmable" waters by 1983 and elimination of pollutant discharges by 1985. Towards achievement of these goals, the CWA mandated:

1) The development and publication by EPA of freshwater and marine water quality criteria which are expressed as constituent concentrations, levels, or narrative statements representing a quality of water that supports a particular, beneficial use. When such criteria are met, the surface water quality will generally protect the designated beneficial use(s) of affected water bodies. EPA's water quality criteria are published as nonenforceable guidance for inclusion in State water quality standards.

The States are required under the CWA to adopt enforceable water quality standards that consist of designated, beneficial uses for all surface waters, water quality criteria to protect the beneficial uses, and an antidegradation policy applicable to very high quality waters. As indicated above, the State water quality standards may include EPA's water quality criteria.

2) promulgation by EPA of directly enforceable treatment technology requirements, such as best practicable technology (BPT) for all types of pollutants and best available technology (BAT) for industrial toxic pollutants, including pretreatment requirements for industrial 'discharges of toxic pollutants to POTWs. EPA has promulgated BAT performance standards for 23 "standard industrial categories." These performance standards establish enforceable effluent limits for a total of 126 toxic pollu-

tants, including all surface organic chemicals 13 metals, asbestos, and cyanide. Covered industries are required to meet applicable performance standards, but the choice of treatment technologies to meet those standards is the prerogative of the affected industrial facilities. Note that these enforceable performance standards dictate minimum treatment requirements that have no direct relationship to the quality of receiving waters. That is, they are not water quality-based treatment requirements.

11. BPT for POTWs has been interpreted by EPA as secondary treatment, that is, removal of BOD and suspended solids down to concentrations in the effluent of 30 mg/l. Most coastal POTWs are now either in compliance with secondary treatment requirements or are on an enforceable schedule to achieve compliance. Section 301(h) variance provision was included in 1977 CWA amendments in response to assertions that POTWs discharging to relatively deep, well-flushed marine waters can protect human health and the marine environment without fully complying with secondary treatment requirements. Evenso, in 1981, section 301(h) was amended to to prohibit permit issuance for sludge discharges to marine waters.

12. EPA has strictly administered the section 301(h) secondary treatment variance provision. For example, of 208 coastal POTWs that have applied for secondary treatment variances, only 73 have received final or tentative approvals. Also, EPA has granted no variances for large POTWs (that is, larger than 5.0 MGD) that discharge to the relatively shallow waters of the Atlantic coastline or to any POTWs that discharge to estuaries. Finally, EPA has granted no variances to POTWs that provide less than primary treatment. Consistent with EPA's strict admistration of section 301(h), Congress shows no inclination to liberalize eligibility for such variance. In fact, the inclusion of more stringent secondary treatment variance provisions in 1986 amendments to the CWA provide clear evidence of Congressional intent on this subject. (Note that, for budget control reasons, the President did not sign the 1986 CWA amendments.)

13. As mentioned above, the CWA mandated industries that discharge directly to surface waters to install BAT technologies to meet EPA promulgated performance standards (effluent guidelines) for toxic pollutants. The CWA also required EPA to establish a comprehensive industrial pretreatment program applicable to an esti-

mated 14,000 industries that discharge to POTW collection systems instead of receiving waters. The industrial pretreatment program is designed to protect POTWs from industrial pollutants that would pass through a POTW untreated or that would interfere with POTW treatment processes. Thereby, the pretreatment program is intended to minimize the concentration of industrial toxic pollutants in POTW effluents and sludges.

14. EPA's BAT standards, as described above, include pretreatment requirements for covered industries. Like the BAT standards, pretreatment requirements are expressed as performance standards (effluent limits) rather than treatment technology requirements. In addition to these national, categorical pretreatment requirements for covered industries, municipalities are authorized to impose local pretreatment requirements on industries that are exempt from EPA's requirements in order to ensure compliance with POTW effluent limits.

15. Protecting the quality of POTW sludges is an important element of current sewage and sludge regulatory activities because, under section 405 of the CWA, as amended in 1977, EPA is also required to promulgate guidelines for POTW sludge management and disposal practices that will actively promote beneficial uses of sludge while maintaining or improving public health and the environment.

16. Whenever technology control requirements, such as POTW secondary treatment and industrial BAT requirements, are not enough to ensure compliance with receiving water quality standards, the CWA authorizes imposition of even more stringent technological controls on M & I pollutant sources. For example, as one element of the Chesapeake Bay (water quality management) Program, major municipal dischargers within the Bay drainage area are required to meet stringent phosphorus removal requirements in addition to secondary treatment. However, under the CWA and as a practical matter, imposition of such "advanced waste treatment requirements" (AWT) on point sources of pollution must be balanced against the need for other pollution control measures, such as best management practices, to control nonpoint pollutant sources.

17. The primary mechanisms for imposing CWA regulatory controls on point sources, such as POTW treated sewage and sludge discharges, are:

> 1) the National Pollutant Discharge Elimination System (NPDES) which is a discharge permit program that establishes effluent limits and

monitoring and reporting requirements on the
permitted dischargers, and

2) pollution control requirements of the POTW
construction grants program which, in effect,
help to ensure that Federal subsidies for POTW
construction achieve the intended pollution
control purposes. The CWA authorized increased appropriations to subsidize POTW construction nationally and to increase the portion of
individual POTW construction costs that may be
funded by CWA construction grants. Currently,
up to 75% of eligible construction costs for
individual POTWs may be funded at a total
annual cost, nationally, of $1.2 billion to
$2.4 billion. Eligibility for these 75% con-
struction grants is tied to a demonstration
that CWA treatment requirements will be met,
including (as stated above) AWT requirements,
if needed, to meet water quality standards.
However, the majority of POTWs in the United
States are required to meet only secondary
treatment requirements.

18. It should be evident from this thumbnail description of current regulatory controls and mechanisms that
water pollution control under the CWA is largely a
technology-based, point source control program by com-
parision with the water quality based pollution control
programs of the "early years". Water pollution control
responsibility, including the regulation of sewage and
sludge discharges to marine waters, resides primarily
at the State and local levels. For example, the CWA
authorizes delegation of permit issuance and enforcement responsibility to States that have approved NPDES
programs, although EPA retains back-up enforcement water
quality standards out to the three mile, territorial
sea limit and to impose specified environmental protection criteria in our decisionmaking on proposed ocean
dumping activities. We are also required to enforce
London Dumping Convention prohibitions and limitations
on specified ocean dumping activities and pollutants.
In addition, waste materials, such as POTW sludge, can
only be dumped within sites that are formally designated
by EPA. Neither ocean dumping nor dumpsite designations
"work" without the other; you cannot dump sludge at a
designated dumpsite without a permit, and permits cannot
be issued for dumping sludge, for example, at a other-
than-designated dumpsites.

19. Conceptually, this decentralization of authority
under the NPDES, construction grant, and pretreatment
programs should encourage consideration of the environmental effects of multiple pollutant sources in decision-

21

making because the states are also directly involved in water quality management planning. Also, the CWA does require consideration of multi-source impacts in NPDES and construction grant decisionmaking. Nevertheless, individual point sources tend to be reg- ulated on a case-by-case basis without detailed consid- eration of other pollution sources. This is because permit issuance and construction grant approvals most often depend upon achievement of tecnology-based treat- ment requirements, BPA and BAT for example, not water quality requirements. This is particuarly true for NCW municipalities because, historically, the quality of NCW "just hasn't been a problem". As will be observed later, this case-by-case decisionmaking may detract from the effectiveness of the total program.

20. As stated earlier, the MPRSA regulates the dumping of materials into open ocean waters. This includes dumping of sewage sludge. Under the MPRSA, it is EPA's policy:

1) to protect the oceans from significant adverse effects of waste disposal. EPA will not allow the oceans to be used as a "cheap" alternative for the disposal of POTW sludges or other wastes;

2) To allow ocean disposal of POTW sludges only when there are no practicable, land-based al- ternatives that have less impact on the total environment; and

3) over the long run, to encourage environmentally beneficial approaches to POTW slude management, also, minimization and recycling of other wastes.

21. As with the CWA, the MPRSA establishes environ- mental criteria for ocean dumping activities and a permit program for imposing those criteria. Specifi- cally, we are required by the MPRSA to enforce State water quality standards out to the three mile, territo- rial sea limit and to impose specified environmental protection criteria in our decisionmaking on proposed ocean dumping activities. We are also required to enforce London Dumping Convention prohibitions and pol- lutants. In addition, waste materials, such as POTW sludge, can only be dumped within sites that are for- mally designated by EPA. Neither ocean dumping nor dumpsite designations "work" without the other; you cannot dump sludge at a designated dumpsite without a permit, and permits cannot be issued for dumping sludge, for example, at a other-than-designated dumpsites.

22. The MPRSA mandates that dumpsites can only be designated by EPA, also ocean dumping permits can be issued only by the Corps of Engineers (COE) for dredged materials or EPA for non-dredged materials. In addition to these EPA and COE responsibilities, the MPRSA requires the National Oceanic and Atmospheric Administration (NOAA) to conduct long term research and monitoring on the effects of ocean disposal activities on the marine environment and living resources. Like the NPDES program and construction grants program, the ocean dumping permit program is administered on a case-by-case basis for individual dumping activities. Also, ocean dumping activies must be consistent with approved coastal zone management plans of affected states.

23. These, then, are the highlights of current pollution control efforts under the CWA and MPRSA as they affect marine disposal of sewage and sludge... stringent point source control programs that are administered primarily by State and local authorities under the CWA and the Federal government under the MPRSA.

24. Under the CWA, marine discharges of inadequately treated sewage and sludge have been reduced substantially. All POTWs that discharge to marine or estuarine waters will comply with minimum treatment requirements within the next several years and, as stated earlier, sludge discharges to marine waters are prohibited. Also, the industrial discharge pretreatment program is already having beneficial effects on the quality of POTW effluents and sludge. When fully implemented, it is expected that industrial pretreatment will reduce discharges of so-called "priority pollutants" to recieving waters by about 75%.

25. Under the MPRSA, sludge dumping quanitities have been fairly stable. Currently, nine municipalities are ocean dumping some seven million tons of sludge annually. However, consistent with the intent of Congress, EPA has refused to allow continued dumping at a shallow, nearshore site within the New York apex because of adverse impacts on affected marine ecosystems and has designated a dumpsite in much deeper waters off the continental shelf. All ocean dumping of POTW sludge will be at the deep water site beginning in 1988.

26. Although these kinds of regulatory programs under the CWA and MPRSA should preclude significant adverse effects from ocean disposal of sewage sludge on a case-by-case basis, NPDES ocean discharge permit issuance and ocean dumpsite designation and permit issuance often do not adequately address the interactive effects of such disposal activities in combination with the

effects of relatively nonregulated pollutant sources, such as combined sewer overflows (CSO), surface runoff from agricultural lands, and uncontrolled discharges from coastal hazardous waste sites.

27. Unfortunately, the combined effects of all pollutant sources, both regulated and nonregulated, are continuing to degrade our NCW. This is because the NCW often act as sinks for waste materials that are inadequately mixed with deep, open-ocean waters. In the United States, the NCW are forced to absorb the wastes of close to 75% of our total population or about 180 million people. Evidence of continuing degradation of our NCW and associated living resources is all too plentiful. We have, for example, documented evidence of five kinds of major problems:

1) widespread toxic contamination of fish and shellfish,

2) eutrophic conditions and hypoxia over thousands of hectares of coastal waters,

3) widespread pathogenic contamination of fish and shellfish as well as thousands of reported cases of gastrointestinal infections of persons consuming pathogen-contaminated fish and shell-fish,

4) loss of, or adverse changes to, thousands of hectares of essential marine habitat each year and at increasing rates of loss, and

5) loss of living marine resources, such as substantial declines in recreational and commercial fisheries, as well as threatened and endangered birds and mammals.

28. Of course, marine disposal of sewage and sludge is only one of several causes of these kinds of problems. Indeed, because of the complex interactions among multiple pollutant sources, it is very difficult to measure, or even reasonably estimate, the relative contributions of multiple pollutant sources to environmental problems of our NCW.

29. Thus, we find ourselves at crossroads. We have committed and are committing large amounts of time and resources to the control of point sources of pollutants, including regulation of the marine disposal of sewage and sludge. And, yet, we see clear evidence of continuing environmental problems in certain NCW. At these cross-

roads in time, decisions are needed that will increase
our commitment towards even more stringent controls on
individual pollutant sources where this is a meaningful
and practicable course of action. We will also need to
broaden the scope of our regulatory efforts to include
controls on other sources of marine environmental pollutants. Balancing the levels of controls on point and
nonpoint sources and on specific and nonspecific pollutants will be a major challenge for us.

LOOKING AHEAD...EPA's DECISION

30. EPA fully recognizes that strict enforcement of
existing regulatory authorities over ocean disposal of
sewage and sludge, as well as other point sources of
waste materials, on a case-by-case basis is not enough
to prevent continuing degradation of our NCW. Until
recently, we have lacked a national policy and commitment that focuses coordinated Federal, State, and local
action on NCW environmental problem solving on a multimedia basis. Now, however, EPA is addressing this
situation through implementation of a NCW Strategic
Planning Initiative that is designed to:

1) redirect selected EPA regulatory activities
   towards the NCW,

2) focus greater attention on developing marine
   water quality and sediment criteria that can
   be used for regulatory decisions affecting
   both coastal and inland activities,

3) develop policies and regulations to control
   CSO and stormwater contributions of pollutants
   to the NCW,

4) revise EPA's five-year research plan to focus
   on NCW needs,

5) use the Chesapeake Bay Program as a case study
   on nonpoint source pollution control. Under
   the Chesapeake Bay Program, Federal, State,
   and local agencies are cooperatively integrating
   water quality based point source controls (for
   example, imposition of AWT requirements as needed to meet water quality standards) with the
   establishment and implementation of voluntary
   best management practices for the control of
   nonpoint pollutant sources within the framework
   of a comprehensive, basinwide water quality
   management plan. This is a multi-media approach
   towards restoring and maintaining high quality

Chesapeake Bay ecosystems through carefully planned reductions of both toxic pollutants and nutrients in critical areas of the Bay.

6) continue to focus National attention on six estuaries (in addition to Chesapeake Bay) that have been designated for coordinated Federal, State, and local action on improvement and maintenance of water quality and related living resources,

7) Coordinate new water quality management activities with other Federal, State, and local agencies, and

8) track our progress over time towards meeting NCW goals and objectives.

31. EPA is also active on a number of related fronts that will affect marine disposal of sewage and sludge, including:

1) revising our MPRSA ocean dumping regulations in response to statutory changes and court decisions and to benefit from a decade of experience administering our existing ocean-dumping regulations,

2) promulgating regulations under section 405 of the CWA that will regulate POTW sludge use and disposal on a multimedia basis. EPA is considering establishment of a combination of sludge pollutant concentration limits and sludge use/disposal best management practices that will provide defined levels of human health and environmental protection. These enforceable section 405 guidelines would be applied nationally to landfilling, land application, incineration, and ocean dumping of POTW sludge. Individual POTWs would have to comply with the guidelines for their selected sludge use/disposal practice(s). It is intended that POTW managers compare the costs and risks of available sludge use and disposal alternatives and select a practicable alternative that complies with the section 405 national guidelines. POWTs could, however, petition for modification of the national guidelines based on local circumstances and a demonstration that human health and the environment would be protected to the EPA-defined levels. Affected municipalities will, thereby, have a multimedia regulatory framework within which they can make cost-effective decisions on the management of their

POTW sludges. These decisions will need to take into consideration the effects of related regulatory activities, such as the industrial

8) revising the hazard ranking system of EPA's Superfund program for "listing" uncontrolled hazardous waste sites to address food-chain contamination risk and environmental protection concerns more adequately. This is a significant action since a number of NCW hazardous waste sites that are adversely affecting marine waters have not qualified for Superfund removal or remedial action under the existing hazard ranking system.

32. In summary, EPA fully recognizes that adequate protection of human health and the marine environment will require continued strict enforcement of existing regulatory authorities over ocean disposal of sewage and sludge, as well as ocean disposal of all other waste materials. This is not enough, however, and we must and will intensify and coordinate Federal, State, and local pollutant control actions on a multimedia basis to halt, and hopefully reverse, the continuing degradation of United States NCW and associated living resources.

33. Therefore, with respect to the hard decisions we need to make at this crossroads of NCW environmental protection, EPA is electing to take a multimedia look at pollutant sources and regulatory alternatives so that the decisions we make on marine disposal of sewage and sludge make good sense in an overall human health and environmental protection perspective.

3. Legal and scientific constraints

P. J. MATTHEWS, PhD, Technical Manager, Norwich Division, Anglian Water

SYNOPSIS. The treatment of sewage and sewage sludge in the sea must be conducted in accordance with external environmental legislation. To do so, and in order to avoid any deleterious public health or environmental effects and to make best use of available financial resources, careful planning control and surveillance is required. This paper describes the practices and problems of a technical manager in an English water authority in meeting those responsibilities.

INTRODUCTION

1.   A traditional maxim — 'where there's muck there's money' is more properly expressed these days as 'where there's muck there's a need for money!'.
This new maxim has been, and will continue to be, evident for sewage disposal.   Public authorities in the United Kingdom with responsibilities for such disposal must ensure that public health and the environment are protected and must do so ·in the most economical way possible.   The need for careful husbanding of resources, particularly available finance, does not mean that "cheap" solutions to the problems of disposal are adopted.   It does mean that priorities have to be allocated to schemes so as to reflect local needs and also that an effective approach must be used to identify the best value for money for investments in individual schemes.

2.   In England and Wales water services are provided by water authorities. These are the major effluent discharges as well as being responsible for pollution control and water quality management. Coastal and estuarial waters are used for the disposal of effluents and for the dispersal of sludges, particularly sewage sludges produced from the treatment of effluents prior to discharge to rivers or to saline waters. The public interest in how these, and other matters of environmental significance are managed is reflected in legislation and in the way in which central government is exercising its statutory function within that

legislation. These legal constraints have, in turn, resulted in developments in technical methodology used to identify the best course of action in any particular location.

LEGAL CONSTRAINTS

3. The Acts of principal interest with respect to water quality are the Water Act 1973, Control of Pollution Act 1974 and the Food and Environment Protection Act 1985. Other Acts such as those relating to oil pollution, pipelines, shipping, public health, and so on, as well as Acts of general aspects such as health and safety, employment etc. are relevant. Laws of tort and contract, for instance, may be of some account but this paper concentrates on the Statute Laws with influence on water quality.

Water Quality Management

4. Under the Water Act the Secretary of State for the Environment and Minister of Agriculture, Fisheries and Food have responsibility for national water policy. The Secretary of State has amongst a number of duties, the duty to secure the effective execution of policy in relation to the treatment and disposal of sewage and other effluents (S.1). This is achieved through water authorities. It is interesting to note that whilst there is a similar duty to restore and maintain the wholesomeness of rivers and other inland waters there is no similar duty for estuaries and coastal waters. However, one of the functions of water authorities was to prepare a plan of action to be taken over a period by way of executing works or securing the execution of works by other persons .... for restoring or maintaining the wholesomeness of rivers and other inland or coastal waters in their area. There has never seemed any explanation for this apparent discrepancy, except that perhaps with the benefit of hindsight no distinction is made now between environmental waters whereas a decade and a half ago, there was. Even this explanation does not bear scrutiny since no distinction was made under the Control of Pollution Act 1974 (COPA). Of course water authorities inherited the powers and responsibilities for the control of estuarial discharges under the 1960 Clean Rivers (Estuaries and Tidal Waters) Act 1960.

5. The COPA was passed in 1974 but there was concern about the potential costs of the provisions of Part II — that dealing with water. Successive governments decided that the implementation would be phased and this is still not complete. However the major tranche of provisions were in place by January 1985.

6.   Under these provisions water authorities have responsibility for water quality management for all environmental waters, now known collectively as 'relevant waters'.   Within the subject of this paper 'controlled waters' means principally the sea within three nautical miles from any point on the coast, measured from low watermark of ordinary spring tides.   'Restricted waters' applies the definition of controlled waters to prescribed areas of tidal rivers and certain vessel moorings.

7.   Under S.31 it is an offence to cause or knowingly permit poisonous noxious or polluting matter to enter the waters with certain exceptions, particularly for trade or sewage effluent discharges and those made in consequence of a licence issued under the Food and Environment Protection Act 1985 (FEPA).

8.   Under S.32 it is an offence to discharge trade or sewage effluent into 'relevant' waters or from land through a pipe into the sea outside 'controlled' waters without consent.   These provisions ensure that the commitments to the Paris convention are satisfied.   The consent is obtained under the provisions of S.34 and associated sections.   The procedures defined in these sections require advertisement of the application for consent except where there is no appreciable effect.   The criteria for defining 'appreciable effect' are given in a DOE circular (ref. 1) although they are of little help when dealing with some waters, particularly coastal waters.   The procedures allow the right of public involvement and in the ultimate event this can result in a Public Enquiry to assist the Secretary of State to determine the conditions of the consent.   The consent may include reasonable conditions to ensure that the quality of the receiving water is protected.   It is an offence not to adhere to those conditions.   The deviation of those conditions is a matter of technical consideration.   These are calculated by water authorities taking account of the nature of the discharge and the needs of the receiving water.   When an application is made by a discharger (mostly of industrial effluent) to a water authority, detailed information on the discharge is provided.   The needs of the receiving water take into account national policy requirements as well as the commitments arising from E.C. Directives.   These will be discussed later.

9.   The dilemma of water authorities being the pollution control authority as well as a discharger (of sewage and sewage effluent) has been overcome by making the Secretary of State the consenting authority for water authority discharges.   The consent conditions are calculated by the water authorities which are scrutinized in detail by the Department of the Environment before the consent is issued.

10. The initial step in the main phase of implementation of COPA in 1983 was to provide exemption from consent for existing discharges which had enjoyed freedom from control prior to July 1984. However exceptions from exemption were provided for discharges which were affected by the requirements of E.C. Directives (ref. 2). Where these were substantially a continuation of discharges made in 1974 the consent was granted by a deeming procedure which allowed the existing circumstances to continue legally. It was anticipated that the full consent would be granted in due course, particularly when the nature of the discharge changed substantially. Other excepted discharges were consented using the full procedure straight away.

11. However the legal situation is about to be made even more complex before it is simplified. The Secretary of State announced in October 1986 that as from October 1987 exemptions from consent for discharges of trade and sewage effluent to controlled waters will be removed (ref. 3). A consent will in effect be deemed to be granted. It is anticipated that all discharges will be consented fully within 5 years. These latter changes have no doubt arisen because of the growing public interest in coastal water quality and international demands, particularly those arising from the European Community that all discharges to marine waters should be controlled as rigorously as those discharging to fresh waters.

12. The requirements of the European Community are expressed in water quality Directives particularly those for bathing water, shellfish water and for the discharge of dangerous substances to the adequate environment, (ref. 4). Another change brought in by COPA was the S.41 Register of consent documentation and monitoring data of all relevant waters and almost all discharges (there are certain exemptions such as Crown exemptions). These Registers are open to inspection by the public and may well become increasingly important for marine waters as interest in bathing waters quickens. The COPA legislation also enables the U.K. to comply with its commitments to the Paris Convention for protection of the sea from land based discharges.

Dispersal of wastes at sea

13. The U.K. is a signatory to the Oslo Convention for the Protection of the North Sea and North East Atlantic. The responsibilities arising from this Convention were discharged by the Dumping at Sea Act 1974 now replaced by the Food and Environment Protection Act 1985 (FEPA).

14. S.5 requires that a licence under S.8 be obtained prior to disposal of waste (i.e. sewage sludge) at sea from vessels. The licence specifies the nature and quantity of the waste, and the area and method of disposal. It is an

offence not to obtain a licence or not to comply with the terms of the licence. For sludges from water authorities the Ministry of Agriculture, Fisheries and Food is the licensing authority. MAFF have chosen to exercise their powers by issuing annual licences. Whilst informal negotiations may indicate a longer term security than that provided by an annual licence, it would be preferable to have a longer period recognised formally. Any argument about the need to respond to unforeseen circumstances can be covered by the power of MAFF to vary or revoke a licence at any time.

15. The issue of licences is accompanied by fees which contribute to the MAFF costs of administration and monitoring. It is noteworthy the water authorities do not have similar powers with respect to their responsibilities. This would require the implementation of S.52 of COPA.

16. It is also of interest to note that FEPA licences are required for marine construction works, such as outfalls, and for tracer chemicals.

17. The E.C. have proposed a Directive for the control of dumping of waste at sea. If implemented this would result in the unjustified and arbitrary reduction of sewage sludge at sea by 50%, and would extend the already established and satisfactory arrangements issuing forth from the Oslo Convention as far as the United Kingdom is concerned. This has not proved a popular proposal and it has been recommended that its adoption should be opposed by the United Kingdom (ref. 5).

18. Under FEPA water authorities have the status of applicant and hence there are not the complications of the dual role which is claimed to exist under COPA. However there is an interacting interface between the requirements of FEPA and of COPA. Care has been taken to avoid double "consents" or offences.

19. In simple terms discharges from vessels are not consentable under COPA but are licensable under FEPA. Discharges of sludge or effluent from outfalls are consentable under COPA but are not licensable under FEPA.

An offence is not committed under S.31 of COPA for pollution arising from an infringement of a FEPA licence. In the event of an uncontrolled deposit of waste at sea, where the waste the source or nature of the waste requires a FEPA licence, this takes precedence over COPA. However a number of wastes are exempt under FEPA, such as fish processing waste and sewage from vessels, and if their deposit within relevant waters causes pollution, an offence under COPA S.31 has been committed.

20. Some effort has been made to explain the legal constraints within a technical context. This is important because to the technical manager in the water industry,

legal constraints are not just a matter of "dusty" theory but do have influence on day to day affairs and demand significant amounts of management time. Whilst the technical aspects are of permanent importance in their own right, the requirements of the law mould the way in which the technical requirements are perceived and satisfied. This has become increasingly evident with the advent of European Community legislature.

TECHNICAL CONSTRAINTS
Economic considerations
        21.   In the United Kingdom the requirements of environmental protection and sound economies are reconciled by the approach of the "Best Practicable Environmental Option".   In simple terms for a given location the enviromental options for waste disposal are evaluated using the best available environmnetal protection criteria.   Such criteria may be derived directly from scientific studies or from statutory requirements and should be selected as being appropriate to the location being considered.   Each of the available options are costed and the cheapest is selected available options are costed and the cheapest is selected as the favoured - i.e. best practicable option (BPEO).   Of course it is not necessary to fully cost all options every time; common sense plays a vital role (ref. 6).   For instance disposal at sea is not a viable option for sewage sludges generated in rural central England.   However, sewage sludges produced in coastal areas may be disposed at sea to agricultural land, to landfill and so on. The "case studies" are rarely simple; flexibility and costs are influenced by factors such as local politics and social attitudes,   inherited   plant   and   the   concept   of regionalisation of sludge management.   Whilst the latter point has been facilitated by the formation of regional authorities in the U.K. (water authorities in England and Wales) this has the paradoxical effect of making the strategic planning more complex yet more flexible; the benefits are however, overwhelming.
        22.   Similar planning can be used for the discharge of sewage from coastal areas where the choice will be between the discharge of screened sewage or effluent from a treatment works.   The point of discharge is an important feature of the favoured option. Where a number of towns are being considered in a scheme, the sewage may be collected in a central point and treated or disposed via a large long outfall.   Alternatively the sewage from each town may be discharged via a local small outfall, although these may still have to be long. Where bacterial standards are important it must be remembered that a long outfall may be necessary even to dispose of treated effluents because these will still contain very high concentrations of

coliform and other pathogenic bacteria. This may, however, be shorter than that required for crude sewage. If local rivers are geographically convenient the option of taking the sewage inland, treating it and discharging to freshwater may be favoured. However, in a number of instances this has now been found to be unsatisfactory because coliform bacteria may survive in the river water to cause bacterial standards in the receiving coastal water to be breached. Of course, in some instances treatment such as chlorination, ozonation or ultra-violet irradiation may be used to improve the bacterial quality of discharges.

Statistics

23. Although BPEO has not been applied as an overt policy for very long it is merely a formal expression of the common sense which has underlaid U.K. practice for a long time and decisions of this type have been taken in some areas. It cannot be denied that in some instances good planning and practice has not prevailed and water authorities still have to remedy unacceptable inheritances as well as coping with constant demand for better and better standards.

24. In 1984/5 83% of 49.8M people in England and Wales were served by sewage treatment with the removal of 1.245M tonnes BOD. 6.65M people were served by sea outfalls. It was reported that there were 686 outfalls as assets of water authorities (ref. 6); however, in a recent report the number of sewage discharges was reported as 499, (there were 118 trade discharges) (ref. 7). Of these only 24 were judged to be significant in terms of substances covered by the E.C. "Dangerous Substances" Directive.

25. The treatment of sewage in England and Wales produces some 1.0M dry tonne of sludge for disposal (after any treatment) of which some 7M wet tonnes were dispersed at sea compared to 8M wet tonnes in 1981 (refs. 5 and 8). By interpolation about 210,000 dry tonnes — some 21% of total production is dispersed in this way from vessels with just over a further 1% being discharged to sea from pipeline. This contrasts with a total national figure of about 30% of 1.2M dry tonnes. There are 10 dispersal grounds but the biggest sources are London and Liverpool.

26. In general terms the water authorities use the dispersive and treatment capacity of the North Sea and Channel for the disposal of about 5-10% of their crude sewage and about 15% of their sewage sludge (ref. 9). To this must be added the use of the Bristol Channel, Irish Sea and Scottish waters.

Criteria for control

27.   The planning of marine water quality is more difficult than that of freshwater but, in general, there are four objectives:

    (i)   protection of public health
    (ii) protection of fish stocks (inc. shellfish)
    (iii) protection of recreational uses
    (iv) protection of the general marine environment and natural habitats

28.   The general objectives expressed above will not necessarily be satisfied at the point of discharge but will be satisfied within an acceptable distance and certainly at the key locations in need of protection – be it bathing water, shellfishery and so on. This is encapsulated in the concept of "mixing zone" for discharges in relation to environmental quality standards (EQS) arising from the 'dangerous substances' Directive (ref. 7). This zone is defined as that area (or strictly speaking volume) where exceeding the EQS is considered acceptable by the authorities concerned and where the boundaries will be monitored to test compliance with the EQS appropriate to the designated environmental quality objective (EQS). The EQO reflects the uses and needs of the receiving water. This zone is not the same as the area over which mixing is taking place, nor is it the same as the area of initial mixing through entrainment. Through the complex movements of water the instantaneous area of mixing can be very different to the mixing zone. The EQS may be expressed as annual average or percentile values (ref. 10). Many factors should be taken into account in determining the acceptable size, particularly the size of zone of biological impact, if any, which may be associated with it (ref. 10).

29.   Mathematical models are vital to the proper derivation of mixing zones. These would be based on dilution characteristics but could be calibrated to take account of physico-chemical changes (ref. 10). WRC have developed criteria based on dilution and the mixing zone concepts to define "significance" of a discharge to estuarial or coastal waters and hence, would be useful in deciding if there is an "appreciable" effect". Although such considerations are perceived as being of particular relevance to trade discharges, a number of sewage discharges contain sufficient quantities of "black and grey list" substances to warrant this attention (ref. 7).

30.   Sewage can affect marine water in ways. First the immediate impact of the waste on the oxygen regime and water quality features such as nutrients and metals.

Second the contribution of flotables. Third the
contribution of pathogenic bacteria. The mixing zones for
these will be different. For instance state of tide,
current and wind direction will have different influences.
Generally for U.K. waters the concerns have been the impact
of bacteria on shellfisheries and bathing waters and of
flotables on bathing waters.

31.   Sewage will always contain organic matter which
could accumulate in the benthos, cause deoxygenation and
excessive concentrations of ammonia (which is a "grey list"
substance). The impact of sewage is usually limited to a
very small mixing zone within immediate vicinity of the
"boil". However, if the receiving water has some hydraulic
restriction where flushing or dilution is less, then the
local impact can be much greater. The most obvious
instances of this are estuaries. The essential point is
that a properly located discharge of sewage allows the
sanitary wastes to be treated using the dispersive power of
the sea. High dilution, flushing and biodegradation using
the enormous oxygen reservoir of sea water is a very
effective form of treatment and the popular image of a
black sludge at the end of an outfall is wrong for the
discharge of properly pretreated sewage via a properly
designed and located long outfall. Such outfalls will not
have an observable zone of biological "problems". This was
recognised as back as the recommendations of the Royal
Commission on Sewage Disposal in 1912 when only screening
and preliminary treatment and no standard was necessary for
dilution greater than 500 (ref. 21).

32.   Far from having deleterious effects on water the
presence of sewage can enrich the eco-system but it is
conceded that without the necessary use of dilution,
dispersion and degradation the local effects may cause
imbalances in the diversities of species. An outfall point
may often be marked by flocks of sea birds, feeding on
sewage solids although this is a picture more associated
with older outfalls. Modern long outfalls will probably
deny the birds any food attractions. Whilst it might be
demonstrated that sewage is attractive to fish etc it has
been argued that it has been a source of fish disease and
genetic malfunctions. This has been argued very strongly
in the case of sludge disposal but there is no evidence
that fish disease has increased in U.K. waters. There may
be problems elsewhere such as in German coastal waters
(ref. 5).

33.   Another potential problem is nutrient enrichment and
subsequent phytoplankton blooms, for example of Phaeocystis
or of the red pigmented dinoflagellates which cause
extensive toxicity problems (refs. 11-14). Such enrichment
may also cause sea weeds to proliferate with consequent
environmental and even navigational problems. Fortunately

there is little if any evidence to link sewage or sewage sludge discharges to repeated or extensive problems of this nature.

34. Usually the most significant feature of the sanitary wastes are the visual and aesthetic effects. Greasey slicks and the presence of floating faecal and other personal wastes are the features which are most readily observed by the public - particularly if beaches are affected. Hence these are the cause of most public criticism. This is often linked mistakenly to health hazards.

35. In Anglian Water the target objective is for its own effluent discharges to coastal waters under normal operation should not cause discolouration or foaming visible from the shore nor cause solid matter to be deposited on the shore; nor cause damage to shellfisheries or conservation areas etc.

36. Microbiological quality is the subject of much attention in the U.K. The two aspects are bathing water and shellfishery water quality. The micro-organisms may either be indicators (of faecal pollution with the general disease hazards that this might imply) or be of interest in their own right. Hence that discussed and studied most widely are the coliforms and salmonellae. If the hazard from total and particularly faecal coliforms is low or non-existent then the potential hazard from other organisms such as enteroviruses will be similarly low. This view is not without critics (ref. 15). There is no evidence that any of the U.K. bathing waters constitute a health hazard but this is a subject of debate within the public area when reference is made to a number of studies in other countries. It is intended to conduct studies in the U.K. during 1987/88.

37. The criteria most frequently referred to are those in E.C. Directives (ref. 4). In the case of shellfisheries, even those where the waters are not designated under the E.C. Directive, the quality of the fish flesh is of paramount importance. Where microbiological contamination of the shellfish flesh renders it unsaleable (in some cases cleansing can be used) and it is likely that the sewage is the cause, it is unlikely that the sewage discharger has any legal liability for any costs or remedial action and this causes debate.

38. The criteria contained in the "bathing water" Directive have been of relevance in the design of outfalls for several years although only 27 waters were designated originally under the Directive in the U.K. (ref. 4, 16). However, the DOE has now changed the definition of bathing water to those waters where there are recognised bathing facilities and there are now 350 (ref. 17). A recent announcement by the Minister for the Environment indicates

that although these waters cannot be designated officially investment will have to be made where necessary over a period of time to ensure water quality consistent with with the Directive criteria.

39. The U.K. practice of marine dispersal of sewage sludge follows the some overall principles as for the discharge of sewage. With good management it is a satisfactory method of sewage sludge disposal (refs. 12, 13). It has a relatively long history extending back almost a century and in assessing effects several studies of well established disposal grounds are available. It should be noted however, that interpolation of observations requires care, since some grounds are subject to contamination from sources additional to sewage sludge. For instance the U.K. is a minority contributor of heavy metals to the North Sea from non-atmospheric sources and sewage sludge from the U.K. contributes less than 5% of the total load (ref. 5).

40. As well as the long established grounds there are a number of areas where dumping is relatively recent and useful comparisons can be made. It is perhaps useful to think of disposal areas in terms of two extensive types - 'accumulating' (i.e. minimum disposal) and dispersing (i.e. maximum dispersal). Accumulating grounds are situated ideally in relatively still water with little current or tidal movement and insufficient depth to prevent wave action from stirring up the bottom. Material dumped would quickly reach the bottom and stay there. Dispersive grounds on the other hand are located ideally in regions of strong and extensive water movement so that dumped material would be carried away and disposed widely. Whilst most areas will be likely to show a considerable range in characteristics and no ground would be totally one type or the other. Most grounds used by water authorities are highly dispersive; a good example is the Barrow Deep in the North Sea, used for many years for the dispersal of sewage sludges from London treatment works. Garroch Head off the west coast of Scotland, which takes sludge from Glasgow is an example of an essentially accumulative ground.

41. In issuing a licence, characteristics of the sludge and dispersal ground are taken into account as well as general matters such as other marine activities which could be interfered with by the disposal operation itself - shipping, mineral extraction, recreation etc. The chemical characteristics of the sludge are important and licences are only issued for sludges with no more than traces of contamination from "black-list" substances: for instance more than 40mg Cd/kg DS or 20mg Hg/kg DS. Bioassay and biological screening tests may be used to provide information on the general toxicity at different levels and against target or indicator organisms.

Planning and operation

42. The application of the laws, concepts and policies described before, requires considerable data and monitoring. For instance in planning a new or altered discharge of sewage to sea it is necessary to determine current movements, dispersion and dilution and treatment capacity of the sea water (ref. 14). Measurement with floats, chemical, radiochemical and microbiological tracers and of existing water quality enable the quality of the receiving water to be predicted at key points such as prescribed bathing areas. The capacity of the sea to "treat" bacterial contamination is expressed as T90 - the time taken to reduce the concentration of a given bacteria by 90%. Values of 34 minutes to 9 hours for total coliforms in daylight have been observed. The principal cause of mortality is solar radiation and hence time of day, time of year and turbidity are important factors. However, it is recommended that a T90 value of 10 hours should be used for design purposes for waters which are not visibly turbid.

43. Physical aspects to the location of an outfall and the design and construction concepts and methods influence the capital and revenue cost of a scheme. The ideal submarine pipe is heavy, strong and flexible (ref. 16). Heavy - to resist drag and uplift forces of current and waves when not buried or before being buried. Strong - to resist the heavy stresses to which it will be subjected during construction and permanent stress during its working life and also to resist unforeseen and accidental damage. Flexible - to take up the bending both from permanent undulations on the sea bed and that induced during construction; the outfall must also be able to take up, by sag, any possible elongated scour hole which may occur and in this respect be long enough to span such scour holes should they occur. Different materials have been used for the pipe - for instance concrete sheathed bitumen coated steel and armoured high density polyethylene. Each will require different laying techniques such as winching and spooling; the pipes are usually buried.

44. The design of the diffusers at the discharge point is important for even distribution and to avoid choking a popular and successful design is the duck-bill port. WRC have provided some role models for pipeline lengths ranging from 0.1 km for treated effluent from 10,000 p.e. to 1.9 km for screened sewage from 1,000,000 p.e. (refs. 11, 14). However local circumstances influence this. For instance the Grimsby outfall serving less than 1M p.e. is 2.6 km x 2m dia. with Aldeburgh less than 5,000 p.e. is served by a pipe 1.3 km x 200mm dia. (refs. 16, 18).

45. Some treatment of the sewage is necessary - either by screening and or maceration or comminution. Where storm

sewage is discharged care must be taken that this is done
in such a way as not to cause localised pollution problems.

46. Equally the dispersal of sludge must be done in such
a manner that problems do not occur. The vessels must be
properly managed and constructed. The larger operations
tend to be conducted by vessels owned by the authorities
and the smaller operations via contractors vessels (refs.
12, 13).

47. The proper planning and "licensing" of any sewage or
sludge discharge must be preceded by proper surveys and
studies. These should embrace not only the physical
characteristics but the marine eco-system. In the case of
sludge dispersal operations the surveys may be conducted by
water authorities as well as MAFF but the latter is vested
with the responsibility to obtain the relevant data - hence
the FEPA licence charges. Equally surveys must continue
once a scheme has been commissioned. The surveys of water,
such as bathing water will take account of local conditions
and such factors as time of year, state of tide, wind
direction, and so on, and samples must be a fair
representation of the prevailing quality and its hazard, if
any of water uses. In the case of sludge dispersal
operations this will be primarily surveillance of the
eco-system not only because of statutory requirements or
even responsible procedure but because it is necessary to
demonstrate to other North Sea riparian states that the
U.K. is being very careful (refs. 5, 12, 13).

48. In the case of sewage outfalls the monitoring has
concentrated on bacterial quality of the coastal waters and
a wider range of determinands for estuarial waters (ref.
14). The advent of European legislation and COPA may well
result in more extensive monitoring programmes.

SOME THOUGHTS FOR THE FUTURE

49. The application of the economic and scientific
principles described earlier are time consuming and the
current and growing interest in saline water quality
management means that more time than ever is expended on
this aspect of the water cycle. In October 1984 William
Waldegrave told the International Conference on the
Protection of the North Sea that, having cleaned up rivers
there was a need to take immediate steps to protect
vulnerable estuaries and coastal waters. The U.K. had
taken and would be taking action in estuaries such as the
Thames, Tyne, Tees and Mersey.

50. In accord with the BPEO policies marine dispersal of
sewage and sludge is a vital feature of providing an
efficient disposal service to the community. Dispersal and
disposal are not really sufficient words because the
activity can be classed fairly as a form of treatment.

51. Whilst BPEO is practised by water authorities and is a feature of government policy, parliamentary committees and Royal Commissions have adopted a cautious supportive attitude. The Royal Commission on Environmental Pollution has published 11 reports since 1971 (ref. 19) and has promoted the idea of BPEO but would like to see unsatisfactory discharges of crude sewage eliminated. However, in its 10th Report it made a very positive statement 'With well designed sewage outfalls we believe that discharge of sewage to sea is not only acceptable but, in many cases, environmentally preferable to alternative methods of disposal'. It has accepted that sewage sludge can continue to be dispersed at sea in a strictly controlled manner as has the House of Lords Committee on European Communities (ref. 5). It is obvious that international and parliamentary pressure is going to continue and this is going to be reflected in changes of government policy, an example is the recent increase in "designated" bathing waters.

52. The public awareness of coastal water quality in particular has increased. The impending consenting procedure for previously exempted discharges will result in a careful examination of the principles and philosophies underlying consents. The preferred option would be for "descriptive" comments in which the original philosophy of the Royal Commission in 1912 is followed. The conditions i.e. location, preliminary treatment, flow, nature, are defined. Numerical consents percentile or otherwise, would not seem appropriate. The services of organisations like WRC with their models will be needed over the coming years.

53. The challenge to the EQO philosophy will continue. The fact that a majority of other E.C. countries have adopted the uniform limit approach has caused some questioning of the U.K. position however technically meritorious it is (ref. 20). Both systems should have as a major feature environmental monitoring and the demands and needs for this to be extended will continue.

54. It is likely that there will be continued pressure to improve the quality of all estuaries not classified as "good" (ref. 23) although before any effort is made it might be appropriate to review the classification system.

55. Whatever the arguments and shortfalls in past practices, current management practices actually work. A classic example is the all embracing example of Tyneside. Objectives were set, the favoured option adopted for a sewage treatment works and a sludge dispersal ground selected and monitored for sludge disposal in the estuary (ref. 6).

56. All this must be accompanied by an agreed re-evaluation of priorities and more importantly availability of capital monies. Without this, other

schemes also deemed high priority will have to be re-scheduled. Nevertheless as capital assets rise so will the revenue expenditure and hence changes as a consequence of financing and increased operational costs. There have been several estimates of costs arising from restrictions and changes. For instance in 1984 it was estimated that if a ban on sewage and sewage sludge dispersal to the North Sea and channel were to be implemented, the capital cost would be of the order of £420M capital and £90M/p.a. operating costs (ref. 6). Costs for the West and Scotland and Northern Ireland would have to be added for national costs. However, the government has announced that the 350 bathing waters recognised by the new DOE criteria will fall within the scope of the E.C. "Bathing Waters" Directive criteria. It expects that the non compliant waters will be cleaned up within the overall agreed limits on total spending.

This announcement has been made even before the two year (1986 and 1987) monitoring period is complete. The period for "clean up" could be as soon as 1997 if the full scope of the Directive is applied. The Water Authorities Association has estimated that this could cost up to £500M in addition to the £300M planned for the next five years in England and Wales.

57. This paper has tried to give a balanced review and recognise the arguments for and against aspects of the issues identified. However, to end on a firm note all the evidence assessed by independent august bodies and by practical experience shows that, with proper planning and control, the sea can be used to treat and dispose of sewage and sewage sludge in a harmless and very cost effective manner. Long may this continue without interference from prejudice and ignorance.

REFERENCES
1. DEPARTMENT OF THE ENVIRONMENT (DOE). Water and the Environment. The Implementation of Part II of the Control of Pollution Act 1974. Circular 17/84 July 1984.
2. HMSO. The Control of Pollution (Exemption of Certain Discharges from Control) Order 1983 S.I. 1983 No. 1182.
3. HMSO. The Control of Pollution (Exemption of Certain Discharges from Control) Variation Order 1986. S.I. 1986: 1623.
4. GARDINER J., MANCE G. United Kingdom Water Quality Standards Arising from European Community Directives. Wat. Res. Centre. Tech. Report. TR204.
5. HOUSE OF LORDS. SELECT COMMITTEE ON THE EUROPEAN COMMUNITIES. 17th Report 1985-86. · Dumping of Waste at Sea. July 1986.
6. WAA. Water Facts. November 1985.
7. WAA. Mixing Zones. November 1986.

8. DOE/National Water Council (NWC). Standing Committee on the Disposal of Sewage Sludge (SCDSS) Sewage Sludge Survey 1980 Data. August 1983.

9. WATER AUTHORITIES ASSOCIATION (WAA). Submission to the International Conference on the Protection of the North Sea. 1984. Reproduced. Eff. Wat. Treat. J. Sept. 1984.

10. ANGLIAN WATER (AWA). Report on the Control of Discharges of Titanium Dioxide Waste to the Humber Estuary. January 1985.

11. WATER RESEARCH CENTRE (WRC). Submission to the International Conference on the Protection of the North Sea. 1984. Sewage Disposal and its Impact on the Marine Environment. 749-M. May 1984.

12. DOE/NWC SCDSS. Sewage Sludge Disposal Data and Reviews of Disposal to Sea. Standing Tech. Report No. 8. January 1978.

13. DOE/NWC SCDSS. Report of the Sub-Committee on the Disposal of Sewage Sludge to Sea. 1975-78. Standing Tech. Report No. 18. July 1979.

14. WRC. Investigations of Sewage Discharges to some British Coastal Waters 1978-1986. TR67, 68, 79, 99, 147, 165, 176, 192, 193, 201, 222, 239.

15. KAY D. McDONALD A. Bathing Water Quality. Euro. Wat. Serv. July 1986. p. 321-328.

16. DAVIS A. L. Flexible outfall at Aldeburgh. Eff. Wat. Treat. J. 1980, 20, (July) 338-342.

17. WHITELAW J., JONES H. E.C. Beach Directive Puts Pressure on Budgets. New Civil Engineer 18-25 December 1986.

18. PAYNTING A. Grimsby outfall - big, far out, deep down. Surveyor 13 Jan. 1983.

19. HMSO. 11 Reports of the Royal Commission for Environmental Pollution, particularly 1st, 3rd, 5th, 10th and 11th. 1971-1986.

20. HOUSE OF LORDS. SELECT COMMITTEE ON THE EUROPEAN COMMUNITIES. 15th Report 1984-85. Dangerous Substances. July 1985.

21. MATTHEWS P.J. Consents - A Philosophy for the late twentieth century. Wat. Pollut. Control 1986, 85, 408-419.

22. DOE/Welsh Office. River quality in England and Wales 1985, HMSO, 1986.

ACKNOWLEDGEMENTS. The author wishes to express his gratitude to Anglian Water for permission to publish and present this paper. Any views expressed are those of the author and not necessarily Anglian Water.

4. Coastal pollution — aesthetics and/or health

J. A. WAKEFIELD, MIEE, DFH, MRSH, Chairman, Coastal Anti-Pollution League Ltd

SYNOPSIS. An outline of the circumstances leading to the formation of the Coastal Anti-Pollution League in 1958. How the medical controversy, together with the difficulty of obtaining information, has delayed action to reduce coastal pollution and how the Golden List of Beaches in England and Wales has influenced public opinion. The case for a public watchdog is made as is the need for better education in the disposal of human waste.

PERSONAL TRAGEDY
1. During the poliomyelitis epidemic of 1957 the author sought advice from his local medical officer of health on whether it was safe for his children to bathe in Stokes Bay, on the Solent, which was known to be polluted by the town's sewage outfall. On being told that there was no foundation for his fears, his wife took their six-year-old daughter in for a bathe, but when they later became surrounded by human faeces they left the water. Fourteen days later their child had all the symptoms of poliomyelitis and within a few days was totally paralysed. She died as a result of the disease three months later. Their son, who had kissed his sister before she became paralysed, contracted the disease 14 days after and fortunately suffered no lasting effect. Their medical specialist showed no hesitation in confirming the author's belief that his daughter had contracted the disease from bathing in sewage polluted sea water.

FORMATION OF THE COASTAL ANTI-POLLUTION LEAGUE
1. The author offered himself for election to Gosport Borough Council and was elected with a large majority. With some difficulty he persuaded the Council to appoint a consulting engineer to advise on what might be done to alleviate the pollution of the Bay. On being told that the outfall should be resited some further distance from the beach, the Council was persuaded to obtain the comparative cost of providing a biological treatment works. During this time the author was constantly being told that nearly every seaside resort disposed of its sewage into the sea without prior treatment and that the Council saw no reason why a comparatively poor town such as

Gosport should do anything different. On being told that the capital cost of a biological treatment works would be twice that of a resited outfall and that the revenue costs would be excessive, the Council chose the cheaper option. It was during this period that local residents persuaded the author to form the Coastal Anti-Pollution League with the following principal objects:-

(a)   To promote the prevention of contamination of the tidal waters and pleasure beaches around England and

(b)   To promote the advancement of education in the science of sewage disposal

MEDICAL RESEARCH COUNCIL MEMORANDUM NO.37   (ref. 1)
1.   Undoubtedly, publication in 1959 of this now famous memorandum was a serious blow to the League, for not only did it deny that there was a serious health risk, it also firmly rejected the idea of adopting a standard based on the measurement of coliform bacteria. It did however imply that there might be some risk from bathing in waters that were "aesthetically revolting".
2.   At the press conference held at the Caxton Hall by the MRC to mark the publication of its Memorandum, the League was able to ask why the beaches that had been investigated by the Council were not named. It was evident from the very high coliform counts shown against some of the numbered beaches that they were "aesthetically revolting". The League maintained the view that the public had a right to know the location of these beaches so that they could avoid using them. On being told that the names could not be revealed because it would harm the commercial interests of the resorts concerned, the League declared that they would somehow find the location of the dirty beaches and inform the public.

THE SEARCH FOR INFORMATION
1.   These circumstances prompted the League to circulate a questionnaire to the then 240 coastal authorities asking them to supply particulars of their beaches and sewage disposal facilities. It was originally intended to publish a "black" list warning the public where not to take their holidays. Not surprisingly many authorities chose to ignore this request for information and those that did comply avoided the truth. On being told by its legal adviser that the League would almost certainly invite proceedings against it in the event that such a list gave any false information, it was decided to publish a "Golden List" and to leave it to the reader to draw his own conclusions. A second letter to those authorities which had ignored the original letter, warning them that their beaches could not be included in the proposed "Golden List" if they continued to withold the necessary information, enabled the League to make inspired guesses as to the state of some of the beaches.

THE GOLDEN LIST OF BEACHES IN ENGLAND AND WALES   (ref. 2)

1. In 1960 the local Member of Parliament helped the League to set up an Advisory Committee in the House of Commons. This prepared the ground for a debate, which was called by Her Majesty's Opposition Party within weeks of the "Golden List of Beaches" being published for the first time. The debate lasted 3½ hours, at the end of which the division went in favour of the government of the day.

ENTER THE LONG SEA OUTFALL

1. Encouraged by the interest shown by the media and more particularly by the general public, the League called a meeting of representatives from all the Local Authorities surrounding the Solent in the belief that there might be a joint solution to the problem which was troubling most of them. It was at that meeting that an engineer, who had been involved in the construction of the Hyperion Outfall in Santa Monica Bay, floated the idea of laying a trunk sewer along the Solent into which bordering authorities might feed their effluent for final disposal in the English Channel well clear of the Isle of Wight. He then gave the author his copy of "Ocean Outfall Design", (ref. 3) which is a comprehensive account of the circumstances leading to the eventual construction of the Hyperion outfall which discharges the sewage from Los Angeles seven miles out into the Pacific Ocean.

2.   Anyone reading this account could not fail to understand the advantages offered by this method of disposal, always bearing in mind the need :-

(a)   To carry out a comprehensive marine survey to establish local current and tidal conditions.

(b)   To obtain a very high dilution factor, which could be achieved by a combination of deep water and multiple diffuser ports. (Hitherto not used in the U.K).

(c)   To ensure that a considerable period of time should elapse before the sewage could return to the shore.

PUBLIC INQUIRIES

1. One of the League's first tasks was to seek a public inquiry into the proposal by Gosport Council to construct a new outfall into the Solent and in this they were successful. Unfortunately MRC Memorandum No.37 was published a few days before the inquiry was due to commence and so, not surprisingly, the League suffered its first defeat.

2. Publicity generated by the Golden List of Beaches and the House of Commons debate led to calls for help from groups in other parts of the country, where sea disposal schemes were being proposed. The League agreed to help the objectors at several public inquiries, but never on the grounds that it was wrong in principle to put sewage into the sea. It was at the

Bognor Regis inquiry that the League first suggested the inclusion of a diffuser and an increase in the length of the proposed outfall, at which point the inspector invited the Council to reconsider its proposal. When the inquiry reopened both suggestions had been incorporated. Schemes at Penzance, Lyme Regis and Bideford were also opposed by the League and were subsequently turned down by the Minister. It is only now that these schemes are being reconsidered and although the consequent delay was regrettable, it has given time for a greater understanding of the factors involved.

CAN THE RISK TO HEALTH BE IGNORED ?
1.   In the beginning the League fully expected the Medical Research Council to be proved wrong, but many years of pains-taking research have not revealed a causal relationship between bathing in sewage polluted sea-water and serious diseases such as poliomyelitis and typhoid. If there ever was a connection it has obviously been minimised by the advent of immunisation.
2.   However, that there is some risk to health cannot be denied. Professor V.J. Cabelli from the United States and others have demonstrated a quantifiable relationship between the mean enterococcus density in marine bathing waters and the swimming associated rate of gastroenteritis.(ref. 4) However this relationship is tenuous and in the League's view need not be taken too seriously. At the worst it might amount to a spoiled holiday from diarrhoea, sickness or perhaps an ear nose or throat infection.
3.   What no-one denies is that it is extremely unpleasant to bathe in human excrement.

GOVERNMENT WORKING PARTY ON SEWAGE DISPOSAL
1.   In 1970 the League was asked by the Working Party on Sewage Disposal headed by Mrs Lena Jaeger to submit its opinions on the disposal of sewage into the sea. Its report "Taken for Granted" endorsed the League's views and in 1972 prompted the Department of the Environment to issue a questionnaire to coastal authorities, which was very similar to that which the League issued in 1959 calling for information about all sewage outfalls discharging into the sea. The League suggested that this report should be drawn up in a form which would be intelligible to the general public and which would enable them to avoid the dirty beaches. In the event the "Report of a Survey of the Discharges of Foul Sewage to the Coastal Waters of England and Wales" (ref. 5) was published in 1973 and turned out to be merely a series of statistics the most significant of which confirmed the League's belief that 85% of our coastal outfalls were unsatisfactory.
2.   Careful study of the report revealed that the Department of the Environment was now in possession of eight-figure map references denoting the point of discharge of all the major coastal outfalls together with a mine of other important information, which the League had for years been trying to obtain. With some difficulty the DOE was persuaded to part

with the map references and population figures. This enabled the League to incorporate this information into a revision of "The Golden List of Beaches in England and Wales" in the manner shown in Figure 1.

NAME OF BEACH	SEWAGE OUTFALLS	POPULATION DISCHARGING FROM OUTFALL	TYPE OF TREATMENT	DISCHARGE POINT RELATIVE TO LOW WATER MARK UNLESS OTHERWISE STATED. DISTANCES GIVEN IN YARDS	E.E.C. STANDARD	REMARKS
1	2	3	4	5	6	7
CORNWALL (cont.)						
ST IVES PORTHMINSTER	1	22 280	D	50 below	*	Sandy Sheltered Safe swimming New outfall planned for 1990
ST LVES PORTHMEAR	–					Sandy Surfing
PORTHERAS COVE	–					Sandy Pedestrian access Swimming dangerous at low tide
SENNEN COVE	1	1 500	A	At LWM	*	Sandy Surfing North of beach dangerous Lifeguards
MILL BAY	–					Sandy at times Swimming safe when calm
PORTHGWARRA	2	(30 (500	D	At LWM		Sandy at low tide
PORTHCURNO	1	1500	A	At LWM		Sandy
LAMORNA COVE	1	?	D	At LWM		Sand and rocks
MOUSEHOLE	1	2 300	A	At LWM		Fishing port
NEWLYN	4	{ 85 3 250 155 6,540	A A A A	At LWM At HWM 50 above 100 below		Fishing port
PENZANCE	3	{ 120 8 050 8 800	A A A	At LWM At LWM At LWM		Sand & shingle
EASTERN GREEN	–					Sandy
MARAZION	2	2 790	A	100 below		Sandy
PERRANUTHNOE	1	1 700	A	At LWM		Sandy
PRUSSIA COVE	–					Shingle
PRAA SANDS	–					Sandy Surfing Lifeguards
PORTHLEVEN	1	10 000	A	At HWM		Flint & pebbles Bathing dangerous
GUNWALLOE COVE	–					Swimming dangerous in rough weather Lifeguards
CHURCH COVE	1	2 000	C	500 below		Sandy
POLDHU COVE	–					Sandy Bathing dangerous at low tide
POLURRIAN COVE	1	2 039	C	At HWM		Sandy
MULLION COVE	–					Sandy Lifeguards
KYNANCE COVE	–					Sandy Safe swimming in calm weather
POLPEOR COVE	–					Sandy Swimming dangerous
HOUSEL BAY	–					Sandy Safe swimming only in calm weather
CHURCH COVE	1	2 190	C	500 below		Rocky fishing cove

Remarks spanning NEWLYN: New sewage disposal scheme proposed for 1990. (not before time)

ABBREVIATED KEY:

A	= No treatment	E	= Primary treatment
B	= Screens	F	= Secondary treatment
C	= Maceration	G	= Other treatment
D	= Tidal tank	*	= Complies with E.E.C. Standard

Figure 1. Page from "The Golden List"

3. Now the League periodically asks the Water Authorities to provide updated information to enable it to revise the List, but unfortunately some Water Authorities are less obliging than others.

4. It seems to the author that, leaving aside the question of cost, the remedy for stopping the pollution of our coasts has been delayed for two principal reasons:-

(a) The endless search for a significant link between the incidence of disease and the degree of pollution, coupled with a failure to recognise that people do not like bathing in sewage.

(b) The witholding of information by national and local authorities.

LEGISLATION

1. In the absence of a standard the League's solicitor advised that there was no possibility of bringing a case against anyone who polluted the sea. The Anglers Co-Operative Association had been successful on numerous occasions, when prosecuting on behalf of the riparian owners of rivers for the loss of fish, but as far as the League could ascertain sewage only served to fatten the fish in the sea. So for 17 years from 1958 until 1975, when the European Commission published its Directive, the League's only effective weapon was the Golden List of Beaches.

THE EUROPEAN DIRECTIVE ON THE QUALITY OF BATHING WATER

1. It is therefore not surprising that in the summer of 1976 the League gave an unqualified welcome to the European Bathing Water Directive (ref. 6) with its bacteriological standard based on total and faecal coliforms. At last the League would be able to take effective action. When asked to appear on the BBC "Nationwide" programme the author said that he hoped Britain would comply, but that if she did not, then the League would be forced to resort to the European Court of Justice.

2. The Directive required member states to nominate waters "where bathing is traditionally practised by a large number of bathers", but the United Kingdom did nothing until 1979 inspite of constant reminders from the League. When, nearly four years after publication, only 27 out of at least 400 beaches were nominated the League despaired of politicians. Neither Brighton nor Blackpool were amongst the 27. When asked by the Royal Commission on Environmental Pollution to give evidence the League drew attention to this absurdity.

3. Fortunately the Water Authorities themselves have adopted a more responsible attitude and within the prevailing financial constraints most have accepted the need to comply with the mandatory requirements given in the annex to the Directive. They are glad to have clear cut parameters against which they can design new discharges.

4. In the League's experience however there are still numerous beaches round our coasts that are at times grossly polluted by sewage, which is discharged from old and often damaged outfalls. Not only is a watchdog body necessary to draw attention to the shortcomings of these old outfalls, such a body is also required to see that new discharges are properly designed and able to meet the European quality standard. Earlier this year the League asked the Minister to call in a £2½ million scheme, which stood no chance of complying with the mandatory standard. The scheme was immediately withdrawn.

BRITAIN'S DESIGNATED BATHING WATERS
1. The results from monitoring the 27 officially nominated bathing waters are given in the table 1.

Table 1. DESIGNATED BATHING WATERS, ENGLAND AND WALES

LOCATION	NAT GRID REF	YEAR					
		1980	1981	1982	1983	1984	1985
Weston-Super-Mare	ST 3154 5993	F	F	F	F		F
Newquay Fistral	SW 7982 6233						
Newquay Towan	SW 8100 6205						
StIves Porthminster	SW 5220 4025					F	
St Ives Porthmear	SW 5150 4103						
Penzance Sennen	SW 3552 2645	F					
Paignton Broadsands	SX 8965 5470						
Paignton Goodrington	SX 8928 5941	F		F		F	
Paignton Sands	SX 8938 6063						
Torquay Torre Abbey	SX 9093 6355						
Torquay Meadfoot	SX 9304 6307						
Torquay Oddicombe	SX 9262 6582						
Weymouth	SY 6810 7950		1/13				
Swanage	SZ 0313 7929		1/20	1/13			
Poole	SZ 0508 8834			1/15			
Bournemouth	SZ 0889 9065				1/14		
Christchurch Avon	SZ 1887 9225	F	1/14				
Ryde (Isle of Wight)	SZ 601 927	F	F	F	F	F	F
Sandown " " "	SZ 599 840					2/24	2/24
Shanklin " " "	SZ 585 811	F	F	F	F	F	1/12
Margate	TR 347 708	F		F			1/14
Southend Westcliff	TQ 8645 8525				1/15		F
Southend Thorpe Bay	TQ 9105 8469	F		1/15	F	F	F
Bridlington South	TA 1805 6610	F	F	F	1/39		
Bridlington North	TA 1895 6718	1/13	F	1/17	F	1/12	
Scarborough South	TA 0461 8860	1/12	F	F	F	1/12	F
Scarborough North	TA 0367 9000		1/11	1/13	1/14		1/18

Key to
symbols: F =   Failure to meet European Bathing Water
                   Standard. i.e. less than 95% of samples
                   analysed were below 10,000 total coliforms or
                   2,000 faecal coliforms
            Fraction =   Number of samples failed/ Number of samples
                   analysed.

2.   By limiting the number of designated bathing waters to 27
Britain has so far evaded the true spirit of the Directive.
However in response to the 10th Report of the Royal Commission
on Environmental Pollution, (ref. 7) the Government has in
December 1985 ordered a major survey of the quality of 350
bathing waters. The waters are to be monitored over two years
according to the regime in the Bathing Water Directive.
Unfortunately the Minister added that "the inclusion of a water
does not necessarily imply that the water will be identified in
terms of the Bathing Water Directive, nor that it will receive
any priority for expenditure". It is known that the European
Commission is preparing a case against the United Kingdom for
failing to nominate waters "where bathing is traditionally
practised by a large number of bathers" and a further statement
from the Government is awaited.
3.   The League knows that some of the 27 nominated bathing
waters, shown in table 1 as complying with the standard, are
sometimes contaminated with gross solids. It is clear that the
Water Authorities also know, for why otherwise would they carry
out remedial works at some of the beaches shown to be in the
clear.

THE IMPLEMENTATION OF THE E E C BATHING WATER DIRECTIVE
1.   The League was the only "non government organisation" to
be invited to a recent seminar in France to discuss the
implementation of the EEC Bathing Water Directive. This was
attended by participants from the European Commission and all
Member States with the exception of Italy, Luxemburg and
Portugal. Five aspects were discussed:-

> (a)   Definition of bathing water
> (b)   Methods of sampling
> (c)   Standards
> (d)   Public information
> (e)   Effect of the Directive in each member country

(a)   the number of bathers is only one criterion among others.
      In practice the different criteria and geographical
      conditions resulted in very different numbers of bathing
      waters being identified:-

      Belgium... 48 areas   Denmark..... 1,374 measuring points
      F R G .... 94 areas   Greece...... 2,750 km
      Luxemburg. 39 areas   Spain....... not available
      Ireland... 6 areas   France...... 3,145 measuring points
      U.K....... 27 areas   Netherlands. 500 (approx.) areas

      The participants were not generally in favour of changing
      the definition, mainly because it is not a good idea to
      change the rules of a game at half time.

(b)   Methods of sampling not only differed from state to state,
      but also between the competent authorities within the

states. No two member states used the same method.
Since, round the British coast, tidal conditions have the
greatest influence on the levels of pollution, it might be
more meaningful to analyse samples collected over a tidal
cycle rather than at high tide at fortnightly intervals.

(c)   There was substantial agreement that faecal coliforms are
      the best general indicator of water quality. The "total"
      coliform count was never the determining factor. However
      there is now scientific evidence of good correlation
      between faecal streptococci concentrations and the
      incidence of infection. An "I" value for this parameter
      should therefore be included in the Directive. Concern
      over the difficulty of meeting the transparency parameter
      was expressed. However this parameter has amenity value
      and in shallow waters, where bathers may dive, it is of
      importance on safety grounds.

(d)   It was generally accepted that there is more public
      information about the problem in the U.K. than in other
      member countries and other countries are now taking an
      interest in the League's work.

(e)   The main effects of the Directive have been;-
      - to establish obligatory standards
      - to intensify the identification of bathing water and
         sampling
      - to encourage investment in bringing water quality up to
         standard.
      - to increase awareness among officials, elected
         representatives and the public of the need to improve
         certain waters.

UNITED KINGDOM LEGISLATION
1.   It was a sad day, when in 1959 our Public Health
Laboratory Service advised the Medical Research Council against
the establishment of a bacteriological standard, for we then
had to wait 16 years for the European Commission to impose
their standard upon us. None of our own legislation has so far
been of the slightest help. Any improvements that have taken
place are the result of pressure from commercial and tourist
interests prompted by the media and the Golden List of Beaches.
No encouragement whatever has come from Government sources. It
might be thought that the phasing in of Part II of the Control
of Pollution Act will help, but that will depend first upon the
establishment of sensible Environmental Quality Standards and
secondly on a massive recruitment of staff to see that they are
observed. The League is not optimistic.

THE NEED FOR A CONSUMER GUIDE AND PUBLIC WATCHDOG
1.   In 1958, when the League was formed, the general public,
if they thought about the subject at all, believed that any
sewage going into the sea must have already received biological

treatment.   If they recognised sewage solids for what they
were, they must either have come from a passing ship or perhaps
the sewage works had broken down.   No authority would ever be
so foolish as to discharge untreated sewage into the sea.
Today the public are better informed, but still not well enough
to know that in the right circumstances the sea itself provides
the fastest and most efficient treatment known.   It is a
strange anomaly, therefore, that the League now finds itself in
the position of persuading would-be-objectors to accept the
idea of marine disposal.

2.   It would seem that there is distrust not only in the
public mind, but also in that of corporate bodies, such as town
councils and chambers of commerce.   For instance the League has
recently been asked by a local authority to monitor its bathing
waters should the water authority decide to construct a long
sea outfall.

3.   Past governments have done little to dispel the mistrust
which undoubtedly exists in the public mind and the exclusion
of the public from water authority meetings has done nothing to
help.   One possible way to regain public confidence might be to
support a body such as the League to enable it to carry out
independent analyses of bathing water samples.

PUBLICITY AND THE MEDIA

1.   Almost every day some section of the media either writes
to or telephones the League for information.   The busiest times
are either in January, when people are booking their holidays,
or in the spring.

2.   Frequently they are looking for a stick with which to beat
the "establishment" and the League is often able to oblige, but
at the same time it is always at pains to explain that the
water authorities have a major task to catch up with the
neglect left by the local authorities when sewage disposal was
their responsibility.

3.   Some sections of the media are more ready than others to
recognise that they have a responsibility to educate as well as
to entertain.   In the author's opinion, for instance, it would
be both entertaining and educational to screen one of the
existing films showing the construction and launch of a long
sea outfall.   Such schemes are major engineering achievements
and would generate interest in the profession.

EDUCATION

1.   Over 2,000 children and their teachers write for help
with their school projects and the League provides them with
posters and an essay describing the basic problems and their
solution.   University graduates also ask for help with their
treatises.   This is a continuous drain on resources, but if
some of the ignorance is to be dispelled it is essential to
continue the work.   So far the League has received no financial
help from the Government.   It derives its meagre income from
membership subscriptions and the sale of the Golden List of
Beaches.

CONCLUSION AND FUTURE OUTLOOK

1. There will always be a need for independent bodies to care for the environment, especially during periods of financial restraint.

2. Such bodies require a modicum of encouragement and financial help from Government sources to enable them to compete with the resources available to the established authorities at public inquiries.

3. The witholding of information about all aspects of the environment generates distrust. It is not only unnecessary, it can lead to serious delay in subsequent reform.

4. Care of the environment attracts foreign tourists and does not involve the importation of foreign goods. Because there is no shortage of labour and all the materials required are indigenous to this Country, Britain should stop dragging her feet and embark on measures to clean up her beaches as a matter of urgency.

REFERENCES

1. Sewage Contamination of Bathing Beaches in England and Wales. Medical Research Council. 1959. H.M. Stationary Office.
2. The Golden List of Beaches in England and Wales. Coastal Anti Pollution League Ltd. 1960. Now in 5th edition.
CAPL, 94 Greenway Lane, Bath, BA2 4LN
3. Ocean Outfall Design. Hyperion Engineers 1957.
828 Figueroa Street, Los Angeles 17, California.
4. Prof. Cabelli V.J. et al. A health effects data base for the derivation of microbial guidelines for municipal sewage effluents. Proceedings of ICE Conference 1981. Thomas Telford Ltd. London
5. Report of a Survey of the Discharges of Foul Sewage to the Coastal Waters of England and Wales. Dept. of the Environment 1973. H.M.Stationary Office
6. Council Directive of 8th December 1975 concerning the quality of bathing water. (76/160/EEC) Official Journal of the European Communities No L 31/1.
7. Royal Commission on Environmental Pollution. Tenth Report.. Chap. IV 86-95. Feb.1984. H.M. Stationary Office.

Discussion on Papers 1-4

DR MATTHEWS, Paper 3
The dual responsibilities of water authorities for sewage disposal and environmental water quality should be emphasized. There are many advantages but also some disadvantages because of conflict of interest. However, the DoE was the consulting authority.

Water management is now a matter of public and political interest and the last few years have seen the growth of environmental pressure goups. Some have been responsible, and in raising political awareness of problems have made the water authorities' tasks of bidding for resources that much easier, but unfortunately there has been irresponsibility and lack of recognition of environmental improvements by others. For instance, it has been claimed that the AIDS virus could live in the sea; this was based on information that the virus could be cultivated under laboratory conditions to produce vaccine, and is a gross distortion of information.

Water authorities are planning ahead for improving coastal water quality and using the principle of BPEO. As a result of changes in COPA all discharges to coastal waters will be consented from October 1987. The initial consents will be simple ones deemed to be consented to discharge but will have to be replaced by full consents as quickly as possible and not later than 1993. I am concerned with the suggestion that the full consents should contain numerical limits for BOD, etc. This can have no real meaning in the circumstances. It would be better to have descriptive consents which specified the location of the discharge pipe and the circumstances of the discharge, e.g. dry weather flow specified and passed through a 5 mm screen.

On the financial side, although the government has set a programme of remedial action by about the year 2000, no extra money is available. Water authorities recognize the need to improve coastal water quality in some areas, for instance, by the construction of long sea outfalls. If more money were available the task could be completed more quickly.

MR WAKEFIELD, Paper 4

A statement from the Government is awaited concerning the number of bathing waters officially notified to the European Commission (EC). A further 352 bathing waters are to be notified, but as far as I know no target date has been given for bringing those waters up to the requisite mandatory standard.

In 1986 the Coastal Anti-Pollution League applied to the Department of the Environment for a grant of $50 000 to set up an independent monitoring service. The plan was to establish three mobile monitoring teams: one each for the west, south and east coasts. Each would consist of a biologist and an assistant and would operate throughout the bathing season. They would be equipped with a mobile home, which would double as a laboratory in the daytime. At least 12 samples from each beach would be collected at hourly intervals and analysed for faecal and total coliforms, using a portable incubator and the necessary sterilizing equipment.

On being asked by the DoE whether we could match a grant of £25 000 to make up the required £50 000, we felt sufficiently encouraged to mention the idea to the water authorities. With one exception they seemed to think it was worth doing, but unfortunately at the last moment we were informed that a grant was not available. We have appealed against the decision on the grounds that confirmation of the official monitoring results carried out by the water authorities would restore public confidence, not only in the efficacy of the long sea outfall, but also in the water authorities themselves. We are awaiting a reply from the minister. As 1987 is the European Year of the Environment, we had been hoping to develop this independent monitoring service as our contribution to it.

The League's crusade has of necessity been a personal effort supported by a loyal committee and some 500 members. Unfortunately we have been unable to persuade anyone to adopt our mantle and so, rather than let our work die with us, we have been obliged to look for some responsible body with whom we could merge. The Marine Conservation Society, with HRH Prince Charles as its President and based at Ross-on-Wye, has volunteered to carry on our work and I shall be recommending our members to transfer their allegiance to them at our forthcoming AGM.

Finally, the Golden List of Beaches has, with the help of the water authorities, been revised. In addition to all the other information it contains, it will indicate which of the 379 bathing waters monitored in 1986 by the water authorities have achieved the mandatory standard given in the annex to the EC bathing water directive.

MR J. K. WATERHOUSE, Sir M. MacDonald and Partners

At the 1976 annual conference of the Institution of Water Pollution Control, Mr Calvert of Taylors, an eminent civil engineer with vast experience in public health engineering, put the case for sea outfalls. He stated that it was inevitable that he would be repeating many of the points which he had made in 1969, 1970 and 1974. At this Conference, Papers based on even more design and operational experience state the same views.

A properly designed sea outfall system will, in the majority of cases, be the best method of treatment for domestic sewage from coastal towns in the UK, providing of course that any potential toxic contaminants from industry are dealt with at source.

There is a real fear that prejudice and ignorance will cause wrong decisions to be made. The lay person has a natural revulsion to the idea of sewage being discharged to the sea and there is a tendency to consider that if a product is cheaper it is in some way inferior. These factors have made and will continue to make it difficult to convince public opinion of the rightness of marine treatment of sewage.

The public are concerned about the health risk, the effect of fish, the aesthetic (recreation) effect, and the effect on marine ecology. All these aspects are dealt with by this Conference. It would be magnificent if the public, and any doubters in the relevant professions, were to be convinced of the rightness and acceptability of marine treatment. (Hopefully, politicians would follow public opinion, especially if the opinion held agreed with the advice of their experts and advisers.)

The EEC approach of laying down criteria for the state of the receiving water is in my opinion to be applauded. Would Mr Mandl please confirm that, at the present time, and with the exception of the Mediterranean, there is no intention of dictating how the sewage has to be treated, and can he give

Table 1. Coliform bacteria
Crude domestic sewage - 100 000 000 per 100 ml

Sewage treatment and disposal to a river	Crude sewage discharged to the sea via a long sea outfall
1. After primary and secondary treatment (90% reduction) 10 000 000 per 100 ml	1. After initial dilution (200 times) at the surface 500 000 per 100 ml
2. After dilution in a river (diluted 8 times) 1 250 000 per 100 ml (diluted 20 times) 500 000 per 100 ml	

his opinion as to the possible future in this respect?

There is one simple fact that should be emphasized to the public. The bacteria count in the river or stream where it crosses a beach, after having received effluent from an inland treatment works near to its mouth, is liable to have a worse bacteria count than the sea water one to two kilometres out to sea immediately over the diffusers of a properly designed long sea outfall (Table 1). Who has not as a child enjoyed playing in dams and suchlike in this exciting part of the beach? The EEC regulations are eminently sensible in this regard, stipulating the bacteria count ceiling of the receiving waters.

In addition, fish love sewage and grow fat on it! Fishermen realize this too.

I would like to ask Dr Davies if his five points of major problems in near coastal waters with well-documented evidence refer to a coastline where there are no other discharges other than inland works with full treatment and properly designed long sea outfalls.

The UK suffers from the 'no sea outfall lobby' pointing out problems arising where there are old short outfalls which have not yet been replaced and then condemning all outfalls.

I would appreciate Dr Davies's views on the necessity and usefulness of treatment other than masceration/screening on domestic sewage before discharge to the sea through long outfalls.

MR P. NEVILLE-JONES, WRc

As an outfall design engineer I welcome the Commission's work on coastal pollution. However, the directives are drawn up for monitoring and are often difficult to apply to design. For an outfall near bathing waters the directive requires that the 95%, 90% and 80% design cases be considered. Near a designated shellfishery the 75% case must also be considered, and for the dangerous substances directive, the annual average conditions apply.

The large number of design cases unnecessarily increases the cost of design and diverts attention from the real design cases. This leads either to increased design costs or poor design. I would like to ask whether the Commission is aware of this problem and whether it intends to rationalize the directives so that they may be used for design as well as monitoring.

MR M. KING, Land & Marine Engineering Ltd

In the light of UK experience, I find it difficult to understand the seemingly extreme position taken by the Environmental Protection Agency (EPA). While acknowledging the wide differences in scale, does Dr Davies not feel that to allow ocean disposal of POTW sludges only when there are no practicable, land-based alternatives is unduly harsh? In general terms, what would Dr Davies regard as practicable,

land-based alternatives with less impact on the total
environment, and if there were such alternatives, how does
he think costs would compare with controlled sea disposal of
sludge?

What evidence has Dr Davies that land-based alternatives
would, in the long term, be less damaging to the total
environment than the controlled disposal of sludge into deep
water - that is to say into well-monitored ocean sites?

I ask these questions as a member of the DoE UK Working
Party on the Disposal of Sewage Sludge into the Sea, and
because there is apparently conflict in experience between
Authors of Papers at this Conference. This, in many ways,
is healthy and thought provoking, but it is perhaps
important to retain an open mind.

There was a time when CAPL was opposed to the discharge of
sewage into the sea. I have always spoken in favour of
marine treatment, under the right conditions. I was among
the first to suggest to the then Ministry of Housing and
Local Government that if it was safe to dispose of trade
effluent, including low level radioactive waste, by long sea
outfalls, then it seemed that it must be safe to use marine
treatment for domestic sewage.

The Bridport long sea outfall in Dorset - built by Land &
Marine in the 1960s - was possibly the first outfall in the
UK specifically designed to discharge municipal waste. I
believe Mr Wakefield opposed that scheme. As it happened
the opposition lost the day, and because in a way this was
in the nature of a test case, that decision was important.
I well recall CAPL's opposition to the Bideford outfall in
North Devon in 1972. Around that time he also opposed a
scheme for Lyme Regis. That was not for an outfall but for
a contact stabilization plant to be constructed on the
foreshore. I seem to recall that Mr Wakefield thought that
the then Borough Council had it all wrong, that it was
nonsense to go for full treatment and that the answer was to
build an outfall! I was in a quandary. In my heart of
hearts I could not agree more, but as chief technical
witness for the promoters, I had to bat on another wicket.
If it is any consolation, South West Water are soon to
construct a long sea outfall at Lyme Regis.

At Bideford, though, Mr Wakefield changed his stance. The
promoters of the joint outfall scheme were told by the
opposition that they should go for inland treatment.
Bideford is still waiting for an answer to a problem which,
to my knowledge, has existed for thirty years.

The Ministry did not approve either scheme, but not
because of the opposition of CAPL. The proposal at Lyme
Regis was technically acceptable: it was rejected on
planning grounds. At Bideford, the outfall scheme was not
approved because the minister felt that more investigation
should be made into the siting of the point of discharge and
into the possible effect on the marine environment in total.

After these early skirmishes, however, CAPL have now given

helpful evidence at many public inquiries and local
meetings. Mr Wakefield personally was of great assistance
to those of us acting for Wessex Water at the time of public
consultation over the Weymouth and Portland Marine Treatment
Scheme.

In the main, I do not quarrel with Mr Wakefield's
conclusions, except to say that I do not believe that
Britain can be rightfully accused of dragging her feet. Mr
Wakefield only has to study the five-year programmes of UK
water authorities to see the massive injection of resources
into cleaning up bathing beaches around our coasts. There
are, in my view, other countries which are dragging their
feet. Britain is vigorous with measures to clean up bathing
waters as a matter of urgency, and water authority capital
programmes provide clear evidence that this is so.

MR G. M. GIBBS, Southern Water

It seems that the US Clean Water Act results in an
egalitarian approach to US sewage effluent standards and for
this reason it would appear (notwithstanding waiver
provisions) that marine discharge effluent standards are
unreasonably severe. In the UK, effluent standards relate
to the nature and extent of the receiving water.

Is there any evidence that the need to incinerate or tip
sludge has a less severe environmental impact than ocean
disposal in the Los Angeles area?

MR A. A. HOULDEN, Welsh Office

I would like to ask Dr Matthews whether he is absolutely
sure that his interpretation of the best practicable
environmental option is or will become the generally
accepted one. I do not remember ever seeing an 'official'
definition but I would guess that his will fit the bill 90%
and more of the time. However, I think that the expression
could have several interpretations, among which would be

   (a) the best scheme environmentally, albeit not the
       cheapest
   (b) the best practical (cheapest) environmental option in
       a wide 'national' sense which, again, might not be
       the cheapest as far as the individual water authority
       is concerned

I really cannot see the need to invent new jargon if all
it is to mean is project appraisal where the options all
have to be environmentally acceptable in view of the
controls in force.

I agree with Dr Matthews that in terms of the normal
discharge consent parameters, BoD etc, the descriptive
consent fits the bill. There is massive dilution available
and there is plenty of oxygen in the sea. The descriptive
consent in the sense of the no sampling requirement would
not, of course, be appropriate in the case of dangerous

substances and bacterial concentrations in environmentally
sensitive adjacent areas. For the latter I think it
madatory that an intensive post-project approval be carried
out to ensure that speedy knowledge is gained as to whether
or not the design objectives have been achieved. After
this, normal routine sampling will always be necessary.

Perhaps the environmental objective concentration could be
included in the consent. A determination of concentrations
at the head works would be inappropriate. This would be a
radical departure from traditional inland water consenting
and it may not be acceptable to our legal colleagues. It
might also be considered sometimes to be impractical if
other discharges, say from rivers, complicate the issue.
However, the effect of these other discharges should have
been assessed in the initial investigations and in most
cases control here will be in the hands of water authority,
i.e. reducing bacterial concentrations in upstream sewage
treatment works effluents if necessary.

I believe that there is a need, particularly in view of
possible privatization, to ensure, in a direct legal way,
that a perhaps reluctant water authority takes remedial
action expeditiously if the design objectives are not met in
practice.

Has Dr Matthews any comments or suggestions to make on
these issues?

We frequently grumble about aspects of present-day life
and wonder what can be done about things. Mr Wakefield's
sheer persistence over so many years is to be admired. In
the section in his Paper dealing with UK legislation Mr
Wakefield states that he is not optimistic that recent
legislation will have the desired effect. In spite of
caviats in official statements I do not agree with him. I
am optimistic. I feel that the nettle has been grasped and
that over the next ten to fifteen years we will see a
dramatic improvement. If this does happen does Mr Wakefield
see a need for the continuation of Coastal Anti-Pollution
League operations? The original aims will, I imagine, have
been achieved. As far as public information is concerned it
is quite likely that holiday towns will include in their
brochures information about what has been done and how clean
their beaches are. This is common practicé in Italy.

The prospective merger with the Marine Conservation
Society and the possibility that an independent monitoring
service will be established, albeit in a small way, seemed
to me to be good ideas, providing such a service was
strictly factual and never sensation-seeking.

MR W. G. G. SNOOK, The Kenny-Snook Association
I have known Mr Wakefield since 1959, when I invited him, in
his capacity as Chairman of the recently formed Coastal
Anti-Pollution League (CAPL), to attend the launching of
Europe's first long submarine pipeline outfall. I hoped
that this new technique could be developed as a means of

providing the solution to the problem of coastal pollution.

Mr Wakefield was quick to appreciate the potential and, subject to cautionary reservations, was prepared to support the technique. He has continued to do so and his courage during the very emotive early days of bigoted opposition was a great comfort and in no small measure contributed to the fact that the technique is now widely accepted by the general public. It is essential that we jealously guard this trust.

The tragic experience which first involved Mr Wakefield in the CAPL was related to the poliomyelitis epidemic. Today we are faced with an even more horrifying epidemic - AIDS.

Apart from sexual contact and direct emplacement in the bloodstream, other possible transmission mechanisms of this disease are still the subject of some debate and research - but with general assurances that the existence of such additional mechanisms is unlikely. It is hoped that such assurances include transmission by mosquito and of enteric and faecal matter.

Extreme caution is necessary, and I strongly believe that marine treatment of sewage utilizing long submarine outfalls and properly designed and located diffuser systems is the most effective way of minimizing the risk of transmission of disease.

The effluent from secondary treatment will still contain five to ten per cent of original enteric bacteria present in the effluent. Residual chlorination of the effluent may well sterilize the effluent but may also produce organohalogens, which are near the top of the list of dangerous substances, the discharge of which is prohibited by an EEC directive of May 1976. (See reference 5 of Paper 1.)

I have always maintained that the effluent from a secondary treatment plant would need to be discharged to sea through a submarine pipeline with a properly designed and located diffuser system to achieve EEC standards. It is most encouraging to see the development and continuity of the WRc as probable world leader in this field, mainly under the auspices of the water authorities.

There was a time, not so long ago, when the academic dominance of design produced some embarrassing results. For example, there was an obsession with achieving port discharges with a Froude number as near unity as possible. This was theoretically and academically correct, but in spite of strong warnings about the outcome, several expensive outfalls suffered blocked diffusers.

I hope that too much reliance is not being placed on the location of diffuser systems from the results of mathematical models, unless these models undergo continuous updating by calibration from field results. I also hope, therefore, that the steering committees directing the work of researchers will continue to be well-endowed with practising engineers.

The ICE Charter defines the profession of a civil engineer as being 'the art of directing the great sources of power in nature for the use and convenience of man'. The sea is probably the greatest source of power in nature. Civil engineers, as a profession, are utilizing its tremendous self-cleansing and purifying properties for the use and convenience of man.

The sea belongs to men of all nations. The fullest international co-operation to maintain a correct balance of scientific and engineering excellence should be continued.

MR I. D. BROWN, Grampian Regional Council

I thought that Greenpeace and Friends of the Earth did not really jump on the environmental bandwagon - they were the bandwagon - or at least the prime movers.

I agree that a separate authority constituting an independent monitoring service for sewer outfalls is highly desirable. Can I commend the structure in Scotland where poacher and gamekeeper have entirely separate identities and do in fact fulfil the role envisaged by Mr Wakefield? As Director of Water Services for Grampian Region, I have on more than one occasion been threatened with prosecution for causing unacceptable discharges to the marine environment and I can assure my English colleagues that there is nothing more likely to spur one into action than the prospect of a spell behind bars?

DR B. DENNESS, Bureau of Applied Sciences Ltd

Mr Wakefield's call for independent monitoring of bathing water quality to correct the present situation where water authorities carry out their own assessment in England and Wales may be reinforced by reference to recent experiences in the City of London. That august financial body has been allowed to police itself. Result? Insider dealing.

Naturally, as professional engineers, scientists and so on, water authority personnel can be expected to avoid such a temptation to abuse their practice of the conflicting roles of poacher and gamekeeper. However, the general public might feel more secure in these times of universal 'glasnost' to which we all strive if bathing water quality monitoring were in independent hands. With proper communication between such an independent organization and the water authorities this need not lead to duplication or any additional cost - merely redirection of resources. Any financial counter-arguments thereby disappear, removing the final reasonable barrier to such a move to sensible openness as already exists in Scotland.

Let us join with Mr Wakefield in his quest for independent monitoring before we experience a pathogenic Chernobyl, which we were also assured (by the combined polluting and monitoring agency) could never happen.

MR P. N. PAUL, John Taylor & Sons

Mr Wakefield stated that the British Government had decided
that over 350 British beaches required cleaning up. This is
not correct. The British Government have decided to
designate another 350 beaches and those beaches will
therefore have to comply with the requirements of the EC
directive. A large number of those beaches already comply.

MR MANDL, Paper 1

In reply to Mr Waterhouse, there is no reason, in my
opinion, to change the present EEC water pollution policy.
It is based on the definition of the water quality
objectives or on the limit values for some toxic pollutants.
As far as microbiological contaminants are concerned, the
quality of receiving waters is laid down in the form of
mandatory requirements but no obligations are spelled out as
to the technical means which could be used to achieve this
quality.

A distinction should be made between directives such as
bathing water or shellfish which fix EQO according to the
designated use of water, and the dangerous substances
directives which aim to protect the aquatic environment
against pollution from toxic, persistent and bioaccumulative
substances.

In the case of persistent substances, a marine outfall
does not remove any pollutant but merely gives a dilution.
This type of disposal would therefore not be appropriate and
treatment of the polluted effluent should be envisaged in
this case.

Where conflicting requirements between two or more
designated use of water directives occur, the most strict
requirement should be applied.

DR MATTHEWS, Paper 3

In reply to Mr Houlden, the interpretation of BPEO given in
my Paper is one used in official publications from the water
industry. The definition used for the disposal of waste to
the aquatic environment means that for each waste there are
a number of options of disposal, each having a different
impact on the environment. For each option there are
environmental criteria which must not be exceeded. Having
determind what needs to be done to protect the environment
in each case, each option can be costed. This takes account
of capital and operational expenditure. The lowest cost
solution is then adopted. The classic example is the choice
of sewage sludge disposal option – to sea, to landfill, to
agricultural land or to incinerate. Without the context of
the subject of the Conference I cannot think of an activity
which would be the BPEO in a natural sense but for the water
authority. The nearest is the possibility of saying that
all sewage sludge had to be utilized in agriculture as a
means of beneficial recycling irrespective of the cost
involved. No one has seriously suggested this yet.

It is true that the commonsense approach underlying BPEO
has been available for many years and underpinned project
appraisal techniques. However, a phrase like BPEO is needed
for political and public presentation. It may be jargon but
the lead has been given by the Royal Commission on
Environmental Pollution and others.

I am pleased that Mr Houlden agrees with me about
descriptive consents. Numerical limits would be necessary
for blacklist substances where the appropriate criteria are
fulfilled. Bacterial standards are not appropriate. A very
important matter is environmental monitoring; it is not
impossible or illegal - just very difficult. In spite of
many debates the inclusion of environmental standards in
consents is very rare, if at all, for the simpler
circumstances in rivers (an example would be a condition in
the consent that the discharge should not cause an EQS to be
exceeded at a specified point).

The problem is, as identified by Mr Houlden, the potential
contribution from other sources. I cannot comment on the
impact of privatization but my proposal for discharge
consents would apply equally to water authorities or new
water companies.

Substantial progress has been made in clearing up coastal
waters. For instance, Great Yarmouth is often the subject
of media interest. It is forgotten, or perhaps overlooked,
that Anglian Water are well into a £43 million plan to clean
up an unsatisfactory inheritance.

Much has been said about independent monitoring but all
effluent and environmental water data are available via the
COPA register and of course there are independent bodies
such as the PHLS and District Council Environmental Health
Departments. Scientists and engineers must work together in
what is going to be a hectic period before the end of the
millenium.

MR WAKEFIELD, Paper 4
In reply to Mr Houlden, the continued operations of the
Coastal Anti-Pollution League will certainly not be required
if his optimism is realized. It is a mystery to me that the
resorts that have clean bathing water have not used the
League's accolade more extensively. Perhaps they will start
doing so soon and base their publicity on the recent
revision of the Golden List.

I am glad he approves of an independent monitoring
service. We should like to establish such a service, if
only to lend more credence to the water authorities, whose
public image has suffered since they stopped the public from
attending their board meetings.

In reply to Mr Snook, I agree with all he says and would
add that the civil engineering profession has done much to
provide us with an economic solution to this difficult
problem. The public would be extremely unwise to reject it
in favour of land-based treatment works, which do not, in

fact, dispense with the need for long sea outfalls.

In reply to Mr King, CAPL has never been totally opposed to the discharge of sewage into the sea. We have only opposed schemes which seemed to us to give insufficient safeguard against the return of sewage to the shore. We had grave doubts about the experimental outfall at Bridport, which we compared with that which the Americans had constructed into Santa Monica Bay at Los Angeles, but we never actively opposed the scheme. We certainly did oppose the outfall which was proposed at Bideford, and the fact that the inspector who took the inquiry did not recommend its acceptance justified our action. It is a great pity that it has taken so long for the water authority to offer an alternative, which in my view is even worse than the original rather short outfall. As for Lyme Regis, I was never opposed to the idea of putting a contact stabilization plant on the beach. In fact I supported it at the public inquiry. It was the Friends of Jane Austen who put paid to that scheme by asserting that if she were alive she would not like her view of the famous Cobb marred by a new sea wall.

In reply to Mr Brown, I stand to be corrected, but I think CAPL was in existence long before either the Friends of the Earth or Greenpeace.

I am ashamed to admit that until recently I had no knowledge of the Scottish system, but I now know enough to realize that it has much to recommend it. Recent contact with the water authorities, which on their own admission have lost some degree of public confidence, leads me to believe that they would welcome an independent monitor. Perhaps our successor, the Marine Conservation Society, will be more successful· in persuading our government to part with the very small sum required.

In reply to Dr Denness, as a chartered engineer myself I have never doubted the integrity of my engineering colleagues or indeed the scientific profession. I also believe the general public would feel more secure if monitoring were in independent hands. Opposition to many of the perfectly adequate schemes would then melt away.

In reply to Mr Paul, I am sorry if I gave the impression that all 350 of our designated beaches needed to be cleaned up. This was certainly not my intention. We, together with the WRc, believe that about one third of our beaches would fail the requirements of the EC directive. This means that action will be required at more than one hundred beaches.

5. Sea outfalls for industrial and urban effluent — French practice

B. QUETIN, Engineer, Department of Applied Mathematics, SOGREAH, France

SYNOPSIS. French legislation concerning the protection of the sea is based on the recommendations of the European Economic Community and major international agreements. Since 1966, administrative bodies have been set up to control pollution and finance has been allocated for this specific purpose. Large-scale studies have been undertaken to ensure better control of the convection, dispersion and degradation of pollutants discharged into the sea, concentrating on the formation of wind-induced currents, turbulence mechanisms and the processes of bacterial disappearance at sea. Numerous sea outfalls have been constructed along the Mediterranean coast. In contrast, on the Atlantic coast, only industrial concerns have opted for such structures, while municipalities have preferred treatment plants.

INTRODUCTION

1. Like all the major industrial nations, France experienced considerable economic growth during the 50s, 60s and 70s, one side-effect being an increase in water pollution. An urgent response was needed in order to protect the country's vital heritage of water resources, and so numerous measures were taken in the legislative, technical and financial fields, with widespread support from the French public. Given the economic importance for France of tourism, shellfish breeding and fisheries, such measures were rapidly extended to the maritime sphere.

2. In dealing with French practice in the field of sea outfalls, three aspects will be discussed, namely the administrative organisation and legislation, technical research and actual schemes, which differ considerably between the Atlantic and Mediterranean coasts.

LEGISLATIVE AND ADMINISTRATIVE BACKGROUND

3. As far as water protection is concerned, the French administrative set-up was radically altered by the innovatory Act of 16th December 1964, which introduced partial decentralisation, giving local authorities certain

powers which had traditionally been in the hands of central Government. Under the terms of this Act

(a) pollution control of surface, subsurface and sea water was made obligatory

(b) pollution control was to be paid for by a system of "fees", based on the principle that those who pollute should pay; this principle is widely accepted by all the member states of the OCED and EEC

(c) the country was divided into six administrative regions, corresponding exactly to the catchments of the major rivers

(d) Catchment Area Committees ("Comités de Bassin") were set up to represent users, vote on fee levels and approve programmes for action

(e) Regional Water Authorities ("Agences Financières de Bassin") were created to provide technical and financial assistance for individuals or organisations contributing to pollution control, and to monitor water quality.

4. As a further measure, the Ministry of the Environment was created in January 1971. One of its functions is to monitor and coordinate the work of the Regional Water Authorities. In practice, they are allowed a great deal of freedom.

5. Fees are paid directly to the Regional Water Authorities by major polluters and communes. In the latter case, the cost may be distributed amongst all users of the sewerage networks, in the form of a specific charge added to the drinking water rate.

6. According to French law, these "fees" do not constitute a tax or duty. Under the terms of the 1964 Act they are intended to encourage those who pollute water to make the investment necessary to control pollution. The legislators also wished to introduce the notion of "economic indifference" with regard to these fees, by basing them on the cost of water treatment. They are thus calculated on the basis of the actual characteristics of the discharged effluent, such as the quantity of suspended matter, BOD, dissolved salts and inhibitory matter.

7. This organisation was accepted by the public at large and 20 years later it has proved its usefulness and efficiency. The Regional Water Authorities have in particular contributed to the construction of numerous treatment plants, aimed at reducing the amounts of suspended matter and BOD.

REGULATIONS

8. Concurrently with the setting up of an administrative organisation and the provision of finance,

French legislation relating to water quality was brought into line with the recommendations of the European Economic Community. Regulations follow two main guidelines

(a) determination of quality objectives
(b) prior authorisation for any outfall or modification to an existing outfall, following a public enquiry.

9. Hence, in the case of sea water, France applies standards relating to bacterial contamination in water used for bathing purposes and in shell-fish products intended for human consumption.

Table 1. Bacteriological standards (germ count in 100 millilitres)

Bathing areas	Total coliforms	imperative standard	10 000
		guidelines	500
Shellfish	Faecal coliforms	wholesome area	300
		unwholesome area	300 to 10 000
		(obligatory refining and with special licence)	

10. More generally, any discharge of solid or liquid substances likely to harm marine fauna and flora is forbidden. Urban effluent must be given at least a minimum amount of treatment before being discharged (primary treatment: grit removal, screening, oil removal, etc.).

11. It should also be noted that France has signed a number of major international agreements concerning pollution of the seas, such as those of Oslo (15th February 1972), London (29th December 1972), Paris (11th June 1974) and Barcelona (16th February 1976).

STUDIES AND RESEARCH CARRIED OUT IN FRANCE

12. The engineer faces a number of questions in dealing with sea outfalls. He must accurately evaluate the movement of masses of sea water which have been contaminated by effluent, taking into account two other phenomena which are superimposed on this propagation effect, namely turbulence, which progressively dilutes the pollution, and, in many cases, chemical and biological transformations.

13.   There are two principal dimensions involved: space and time, and it must be admitted that the greater these variables, the less it is possible technically to control the phenomena in question. It is difficult to predict how a large mass of water will behave beyond 8-10 tidal cycles, owing to drift currents. It is extremely difficult, and beyond the capability of the hydraulics specialist, to foresee the long-term consequences on fauna and flora.

14.   Any discharge of effluent into the sea always includes an initial phase where flow is influenced by the characteristics of the outfall and effluent themselves, i.e. emission velocity and difference in density between the effluent and receiving medium. This type of flow, known as the jet or plume (very similar to smoke coming out of a chimney) has been subjected to tests in most hydraulics laboratories. The calculation methods involved are thus well understood, but they call for the use of computers. In a few simple cases, however, calculation nomographs may be drawn up. Those proposed in France have recently been published (ref. 1).

15.   With the exception of certain types of industrial waste, discharged effluent is lighter than sea water and will rise to the surface if the sea water has uniform density. However, this is not always the case. The effluent's floatability may be cancelled out and the contaminated water blocked at an intermediate level.

16.   There are two possible causes for stratification in sea water: either fresh water flowing in from large rivers, or heating of the upper layers by sunlight, of which thermoclines are the most common manifestation. These occur in many lakes and seas, depending on climatic conditions, and notably in the Mediterranean. The seasonal thermocline is located near 30 m depth, which is the practical limit of penetration of sunlight and heat.

17.   Densimetric stratification in seawater poses a particular problem along the French Côte d'Azur, where the bed drops rapidly. A compromise must be found so that the outfall point is located sufficiently far from the shore and beaches without being in water which is so deep that the thermocline prevents the plume from rising.

18.   There are indeed significant drawbacks when the plume is trapped in this way, for example the currents are weaker and no longer ensure rapid dilution of the effluent, and, in particular, there is not enough dissolved oxygen and light to complete the natural self-purification process. Indeed, in Cannes Bay, in 1975, serious algal bloom was noted below the thermocline, proof of a major accumulation of nutrients discharged by an urban effluent outfall.

19.   Thermal stratification depends on the third and fourth decimal place of the density value. This means that water temperature and salinity must be measured with

extreme precision. Such measurements cannot be improvised and experienced oceanographers must be called upon.

20. Jet calculations are vital, as the turbulence in this type of flow enables rapid dilution of effluent near the outfall point. This is extremely important in the case of certain industrial effluents containing high concentrations of harmful substances. On one occasion, SOGREAH designed an industrial outfall with 100 orifices, enabling effluent to be diluted about 200 times at a distance of only 15 m from the point of discharge.

21. Convection and diffusion of contaminants is the second stage of effluent dispersion. Major efforts have been made in France to determine the physical phenomena involved and to develop satisfactory computation tools. As regards the former, two aspects have been studied, namely wind-induced currents and diffusion due to turbulence.

22. Pollution problems become critical when marine currents are weak and no longer ensure rapid dispersion. Currents of this type are due to tide action, but also to winds. It is difficult to evaluate the proportion of current induced by the wind, and all attempts based on measurements and wind-current correlations are doomed to failure. There are at least two reasons for this. Firstly, wind entrains water as a result of surface friction and this movement is transmitted to the body of fluid by internal shear. There is thus a rapid decrease in the speed as a function of depth. The second difficulty is due to the inertia of the body of water. Mathematical models have shown that it takes several hours for a body of water to be brought into movement, but ten times as long for this movement to be dampened when the wind stops. Measurements at sea cannot be relied upon, for this reason.

23. The classic theory established in 1905 by Ekman states that, as a result of the earth's rotation and the Coriolis force, the velocity vector rotates with depth. However, Ekman's theory is based on the assumption of constant turbulent viscosity. This is acceptable in infinitely deep sea but inadequate in inshore waters.

24. Thanks to the progress made in modelling turbulence phenomena, it has been possible to develop specific mathematical models which give a more accurate idea of wind-induced flow. At the same time, these models provide data on the vertical diffusion of pollution. A number of works have been published on these subjects and are listed in the bibliography at the end of this report.

25. Flow in the sea is always turbulent, insofar as there differences in velocity which may cause set up swirling horizontal and vertical movement. Masses of water may therefore be moved in different directions from that of the general flow, thus helping to mix and disperse the contaminants. The eddies are random and difficult to predict, even with careful model simulation. A hydraulic

scale model cannot reproduce both large areas of sea and small eddies at one and the same time.

26. Average flow is therefore reproduced on models, while the effect of diffusion is introduced by an approximate additional term. However, numerical coefficients which can only be determined by experience must also be introduced. Various measurement campaigns have therefore been carried out at different points along the French coast to ascertain orders of magnitude, check those theories which appear to be valid and finalise strict measurement procedures.

27. Complex mathematical models are required to carry out detailed computations of the tide- and wind-induced velocity field at sea. In France, the necessary investment has been made jointly by the Laboratoire National d'Hydraulique (run by Electricité de France) and SOGREAH. The traditional finite difference methods for solving differential equations had to be significantly modified, as they are not sufficiently precise to deal with pollution problems. This is because the discretisation of non-linear terms introduces a systematic error of computation which is greater than the diffusion effect. More sophisticated computation techniques, referred to as fractional step techniques, are required.

28. To be used in design office computations, such a model must solve a number of connected problems, such as the definition of boundary conditions and areas which emerge at low tide. The particular features of wind-induced flow are also introduced by means of a special formulation derived from the particular vertical integral for the corresponding velocity profile (ref. 2).

29. A second mathematical model complements the first. This computes the trajectories of masses of contaminated water and the concentration fields. Two procedures connected with the actual constraints of numerical computations are used. Every model involves the use of a computational grid whose minimum dimension is determined by the cost of the computation being run. However, this minimum dimension constitutes a "filter" and excludes the representation of phenomena at a smaller scale; this is indeed often the case with areas near the outfall point, which receive the initial pollution. For this reason, whenever the areas in question are smaller than several meshes, they are "convected" as a function of the trajectories calculated and "diffused" by integral solutions of the diffusion equations. When the contaminated area is reasonably extensive, a second procedure is used, involving a fractional step method as in the case of velocity calculations.

30. Sea outfall studies performed in France have also dealt with the difficult problem of "bacterial decrease", which ensures the natural self-purification of urban

effluent. This non-conservative process is quite widely
accepted and a number of works have been published on the
subject throughout the world (ref. 5). Extensive
measurement campaigns have been carried out in France along
the Atlantic and Mediterranean coasts in order to evaluate
the bacteria disappearance rate. As a result, the
disappearance rates have been characterised by two
parameters instead of a single one, namely the mean value
and the standard deviation of what is commonly referred to
as the "T90", i.e. the time required to reduce the initial
number of bacteria by 90% (excluding all diffusion
effects).

SCHEMES IMPLEMENTED

31. The idea of treating all effluent before discharge
appeals to the reason. It is supported by the public and by
the French administrative structure. However, it is easy to
forget that the techniques used at present to treat urban
effluent are not effective enough to protect the
environment. When used, chlorination reduces the bacteria
content of urban waste water 100-1000 times, whereas 10^5
would be required to meet bathing-water standards and 10^7
for shellfish farming. Only natural ponding is this
effective, but it is impossible to find the 10 m^2 per
inhabitant required for this technique in the vicinity of
large towns along the coast, and especially in rugged areas
such as the Côte d'Azur and Brittany.

32. To be realistic, the area required should be "taken"
from the sea itself, by means of a sufficiently long
outfall, usually of the order of a kilometre.

33. In France, submarine outfalls have been constructed
mainly in the Mediterranean, perhaps because the area's
tourist trade called for quality water, and certainly
because public funding was therefore made available for
investments.

34. In contrast, along the Atlantic coast, the policy
has been to construct treatment plants. However, there are
still problems connected with the bacteriological quality
of the water. Industrial outfalls exist at several
locations along the Channel and Atlantic coasts.

35. The forces exerted by swell and waves can only be
withstood by massive structures. Large submarine outfalls
in France are constructed with welded steel pipes which are
lined and sheathed with reinforced concrete. The most
commonly used laying technique involves pulling the pipes
along the sea bed. They are kept empty to limit the amount
of friction and traction forces.

36. Pulling operations for large outfalls are very
costly and can be compromised by bad weather. Success
therefore depends on rapid implementation of the work and
attention to every detail of site organisation. An example
of what may otherwise happen was seen at Nice.

MARINE TREATMENT OF SEWAGE AND SLUDGE

Table 2. Characteristics of some large outfalls constructed in France

	Urban effluent				Industrial effluent	
	Nice	Cannes	Menton	Cassidaigne	La Salie	Le Havre
Year of construction	1980	1972	1976	1967	1972	1980
Outside diameter (mm)	1810	1740	800	300	1200	500
Length (m)	1150	1100	1200	8000	800	1300
Material used	Concrete with steel cylinder	Concrete with steel cylinder	Concrete with steel cylinder	Steel	Steel	Steel
Maximum depth (m)	100	85	45	350	10	15
Observations					(wharf)	

37. After the outfall pipe had been pulled into position on the sea bed, it was to be flooded by blasting out the offshore end. The pipe filled rapidly and the pressure exerted on the ground changed just as quickly. It seems likely that the sand and mud forming the bed became fluid and this momentarily reduced the friction coefficient which was restraining the pipe. Longitudinal forces along the pipe centre line rose to the point where the temporary cables holding the pipe snapped and the entire outfall slid forward some 150 m into the sea.

38. I should like to close by recommending that civil works specialists read the publications by R. A. Grace quoted in the references.

REFERENCES

1. QUETIN B. and DE ROUVILLE M. Submarine sewer outfalls: a design manual. Pergamon Press - Marine Pollution Bulletin, vol. 17, no. 4, April 1986

2. HAMM L., QUETIN B. and DE ROUVILLE M. Two-dimensional modelling of wind-induced currents in coastal and harbour areas. International Conference on Numerical and Hydraulic Modelling of Ports and Harbours, Birmingham, 23-25 April 1985, Paper A2, BHRA.

3. GAUTHIER M. and QUETIN B. Modèles mathématiques de calcul des écoulements induits par le vent. International Association for Hydraulic Research, Congress of Baden-Baden, Subject B.a, 1977.

4. QUETIN B. Mathematical model of turbulence simulation: wind-induced destruction of density stratification. International Association for Hydraulic Research, Congress of Moscow, Subject B.a, 1983.

5. GAMESON, A.L.H. et al. Discharge of sewage from sea outfalls. Acts of Symposium of London, Pergamon Press, 1974.

6. DE ROUVILLE M. and QUETIN B. Rejet en mer: dissipation des bactéries. Techniques et Sciences Municipales. L'eau, Januar-Februar 1983, 78th year, nos. 1-2.

7. GRACE R.A. Marine outfall systems. Prentice Hall, Englewood Cliffs, New Jersey, 1978.

8. GRACE R.A. Sea outfalls: a review of failure, damage and impairment mechanisms. Proceedings of the Institution of Civil Engineers. Part 1, 1985, 77, Februar 1978, 137-152.

6. Marine disposal of sewage and sludge — Australian practice

M. W. WHYTE, BE, Director of Support, Metropolitan Water Sewerage and Drainage Board, Sydney

SYNOPSIS. Almost all Australia's large cities are situated along the coastline and take advantage of the assimilative capacity of the ocean for sewage disposal. Increasing incidence of marine and beach pollution has led to the need for upgrading treatment and disposal arrangements. The paper provides an insight into how the major sewerage authorities have approached their individual problems. Sydney, in particular, is currently undertaking the largest scheme of all, involving the construction of extended outfalls from all three of its major primary treatment plants. The paper provides background information on these outfalls and construction progress.

INTRODUCTION

1. A report titled "Engineering Offshore - Engineering Implications of an Australian 200 Mile Exclusive Economic Zone", published almost 2 years ago, states that waste disposal and pollution of the 200 mile zone is likely to be one of the main issues in the management of an exclusive economic zone.

2. In a subsequent article (ref. 1), the convener of the working party which prepared the report, Jon Hinwood, comments on the sources of wastes which pollute offshore waters. The sea's potential capacity to absorb wastes is not being used and more knowledge is required as to what can and cannot be stored in the sea. The idea of saving costs and reducing risks by storing certain wastes in the sea should not be dismissed on emotional grounds but should be carefully studied, he says.

3. Every day, Australians produce 5000 ML of liquid wastes which, after treatment, are discharged to streams, ground water and ultimately to the sea.

4. The report concluded that over the next 25 years the principal risk of pollution offshore is from a number of onshore sources contributing to a confined or limited body of water. These bodies may include Newcastle-Sydney-Port Kembla, Bass Strait, Fremantle-Perth-Kwinana and the South Australian gulfs. The Law of the Sea gives Australia the responsibility of and authority to manage these problem areas and to control pollution within them.

5.   The report provides estimates of the total future
discharges to Australian waters (Table 1).   In looking at these
apparently high figures, it is worth remembering that the water
in a strip of the Exclusive Economic Zone 100 km long has a
typical mass of about a million million tonnes, so that in
general the change in concentrations in the offshore zone due
to waste discharge will be very small.

Table 1.   Discharges of nutrients and trace metals in
domestic and industrial sewage.

Material	Concentration mg/L	Discharge in year given - tonne/year			
		1980	1990	2000	2005
Total N	60	80000	100000	120000	130000
Total P	9	12000	15000	18000	20000
Cadmium	0.001	1.3	1.7	2.0	2.2
Chromium	0.1	130	170	200	220
Copper	0.15	200	250	300	330
Lead	0.03	40	50	60	70
Mercury	0.0005	0.7	0.8	1.0	1.1
Nickel	0.05	70	80	100	110
Zinc	0.3	400	500	600	700

BACKGROUND
6.   In Australia, there is a commonly held view that sewerage
authorities should reuse sewage rather than waste it to the
ocean.   This no doubt derives from the fact that Australia is a
dry arid country.   Opposition to ocean disposal comes also from
long standing problems of beach pollution caused by sewage
discharges.   What the public fails to appreciate is that reuse
on a major scale is not always economically viable with today's
technology.
7.   Small coastal communities often employ secondary
treatment irrespective of whether effluent is reused or
disposed of to the sea.   Exceptions are where there is little
recreational use of ocean waters in the proximity of the ocean
outfall.
8.   This paper focuses on the major coastal outfalls,
particularly those which are the responsibility of the major
urban sewerage authorities.
9.   All capital cities, other than Canberra, are located
along the coastal strip.   Excepting for Brisbane which
discharges secondary effluent to a river leading to the sea,
all make use of ocean waters for direct discharge of sewage
effluent.   The degree of treatment provided varies according to
the population served, the proximity of outfalls to and the
public use made of beaches and, of course, the local marine

environment. As one might expect, the strategy for systems serving large populations is either secondary treatment with near-shore discharge or primary treatment (or less) with long outfalls. Reuse of part of the effluent is practised where viable.

10. It is only recently that some capital cities have been able to commit funds towards completion of their long term sewerage strategies. During the Second World War little capital was available for water and sewerage infrastructure. The post-war boom and continued resource limitations led to large sewerage backlogs in most capital cities, funding available being engaged principally for augmentation of water supply.

11. Public pressure in the 1960's led to increased funding and enabled catch-up of the sewerage backlog in earnest. The emphasis was on sewerage connections rather than satisfactory disposal of sewage effluent and the inevitable result was increasing incidence of beach pollution.

12. The 1970's saw a move of funds towards sewage treatment. Environmental protection authorities were established in most states, reinforcing what sewerage authorities needed to do and even more importantly, causing the progressive elimination of numerous troublesome industrial discharges. Many were, of course, diverted to sewerage systems, often exacerbating the beach pollution problem.

13. Most capital cities have recently completed or are now entering the final stage of their ocean disposal strategy and this will soon lead to the elimination of current coastal pollution problems of sewerage origin and enhance the marine environment.

14. It is interesting to note that all cities have adopted a similar philosophy for ocean disposal where long outfalls have been recognised as needed ultimately. In almost all cases treatment facilities (generally primary or less) have been established first to assure the integrity of the outfall to be constructed later.

INFLUENCE OF COASTAL ENVIRONMENTS ON SEWERAGE STRATEGY

15. Each capital city has a different coastal environment and this has led to differences in the strategies for ocean disposal of sewage and sludge. Table 2 illustrates this and provides information on projects proposed or underway. Figure 1 shows outfall locations.

16. Melbourne, Victoria, is situated on Port Phillip Bay. Effluent from two of Melbourne's three major sewerage systems is discharged to the Bay. Because of questions on assimilative capacity of the bay, a Government decision, influenced by community attitudes rather than scientific need, resulted in secondary effluent from the newly established third system being taken some 56 km for shoreline discharge to Bass Strait. Rural property owners en route are being encouraged to draw effluent from the cross-country outfall for irrigation purposes.

Figure 1. Location of major sewage outfalls referred to in the paper.

17.   The nearby city of Geelong discharges raw sewage to the ocean, remote from the city. Increasing beach pollution problems have led to a proposal involving the milliscreening of raw sewage, grit removal and a 1200m x 1.35m diameter outfall. This work is currently in the construction stage.

18.   Adelaide, South Australia, is situated towards the head of St. Vincent's Gulf. Local factors have led to large scale reuse of sewage effluent, principally for watering of recreation areas and the airport and for irrigating pastures, vines, flowers and selected vegetable crops. Surplus secondary effluent is discharged to the Gulf.

19.   By contrast, raw sewage from Mt. Gambier is conveyed 30km cross country to ocean waters. Despite the remoteness of the outfall, the scale of discharge is leading to unacceptable bacterial contamination and deposition of sewage debris on a popular beach. Secondary treatment is proposed.

20.   Perth, Western Australia, is located on Cockburn Sound which behaves as a "tidal lake" for much of the time and consequently effluent is retained for long periods. Eutrification is well advanced and identified as caused by nutrients from industries and a large primary sewage treatment plant. Two smaller secondary plants discharge via outfalls to the Indian Ocean. The primary plant's discharge point has recently been moved to Cape Peron (ref. 2) which provides a well-flushed ocean discharge point suitable for dispersal of primary effluent.

Table 2. Major Outfalls around the Australian Coast.

LOCATION	POPULATION EQUIVALENT (INCL. INDUST.)	FLOW (ML/D) ADWF	FLOW (ML/D) DESIGN	EXISTING ARRANGEMENTS PLANT	EXISTING ARRANGEMENTS EFFLUENT	EXISTING ARRANGEMENTS SLUDGE	PROPOSED CONSTRUCTION PLANT	PROPOSED CONSTRUCTION EFFLUENT OUTFALL	PROPOSED CONSTRUCTION SLUDGE	COST (A$M)
Sydney:										
Malabar	1 500 000	490	1 550	Primary	Ocean at Cliff Base	Digested (scr) to ocean	Improv. Scum removal	3.5m dia x 4km x 80 m deep	To new outfall (Dig. screened)	94
Bondi	600 000	165	570	Primary	Ocean at Cliff Base	Digested to ocean	Pumping and augmentation	2.25m dia x 2km x 60 m deep		108
North Head	1 200 000	350	1 050	High Rate Primary	Ocean at Cliff Base	Incinerated		3.5m dia x 4km x 60 m deep	(Continued incineration)	92
Cronulla	200 000	50	200	Primary	Ocean at Cliff Base	Land Disposal	-	-	-	-
Newcastle:										
Burwood Bch	220 000	52	500	Screening	Near Shore	(not separated)	Secondary	Shoreline	Pumped to ocean	?
Melbourne:										
Carrum (SEPP)	1 200 000	290	700	Secondary	Edge of Wave-cut platform (some reused)	Land Disposal	-	-	-	-
Geelong:										
Black Rock	175 000			Commintion	Ocean at Shoreline	(Not separated)	Milliscreens & Degritting	1.35m dia x1.2km x 15 m deep	(not separated)	29
Adelaide:										
Bolivar	1 300 000			Secondary	Irrigation, Surplus to coastal swamp	Fertiliser	-	-	-	-
Glenelg	300 000			Secondary	Irrig, surplus to ocean	0.15m dia x3.5km x 12m deep ocean	-	-	-	-
Christies Bch	110 000			Secondary	" " "	Fertiliser	-	-	-	-
Pt. Adelaide	150 000			Secondary	Waterway	0.2m dia x 4km x 11m deep ocean	-	-	-	-
Mt. Gambier:										
Finger Point	200 000			Raw Sewage	0.4m dia x100m x 2 m deep	(Not separated)	Secondary	Shoreline	Land	-
Perth:										
Subiaco	300 000	50		Secondary	1.1 km (ocean)	Land	-	-	-	-
Beenyup	600 000	70		Secondary	1.6 km (ocean)	Land	-	-	-	-
Woodman Pt.	420 000	125	250	Primary	1.8m x 20m deep to sound	Land	-	-	1.4m dia x4.2km x 20m deep ocean	15 (completed)

21.   Sydney enjoys the best flushed coastal waters of all
Australian capital cities, the area being renowned for high
energy.   Also, the ocean floor falls relatively steeply to deep
water, a depth of 60 metres being reached within two to three
kilometres of the shore.   Consequently deep water discharge of
primary effluent through long outfalls is appropriate to solve
Sydney's long standing beach pollution problems.

22.   Of Australia's larger coastal cities, only Sydney,
Adelaide, Geelong and Newcastle discharge sludge to the ocean
on a large scale as will be discussed below.

SLUDGE DISPOSAL
23.   Sludge disposal practice varies according to local
circumstances.

24.   Few commercial ventures have been successful for sludge
reuse on a large scale.   Successes include the sale of most of
Adelaide's sludge to a fertiliser factory which uses it as a
filler after sterilising.   Some authorities are moving towards
marketing waste products and this is likely to lead to further
commercial ventures.

25.   Land disposal is practised to a degree in all cities.
The highly built-up nature of the areas adjoining Sydney's
three major coastal plants would make land disposal of their
sludges a costly proposition.

26.   Adelaide operates two 4 km long sludge outfalls to St.
Vincent's Gulf.   Monitoring has revealed that stable conditions
have been reached within 10 years, the principal impact being
the degradation of seagrasses over a limited area.   A firm
trade waste control policy is central to the continued success
of discharge of screened digested sludge to the Gulf.

27.   Sewage sludge has been disposed of off shore from Sydney
for over 90 years, having been part of raw sewage discharge for
most of this period.   Despite this long period of near-shore
sludge disposal, no significant adverse environment effects
have been identified in surveys conducted since 1972.
Observations in areas directly influenced by existing
discharges revealed that the nature of benthic communities was
influenced by depth of water and the level of suspended solids
in the water column.   The relative proportion of plants and
their associated herbivores decreased with increasing depth and
sediment load as these reduced the amount of light reaching the
seabed.   There was a compensating increase in the abundance of
suspension feeding animals and their associated carnivores.

28.   Small deposits of sludge reported from time to time
could not be relocated on subsequent searching, confirming
theoretical deductions from oceanographic surveys that sea
floor velocities are frequently of sufficient magnitude to
resuspend sludge materials not assimilated.

SYDNEY'S MAJOR SEWER OUTFALLS

History
29.   When commissioned in 1889, Sydney's first two major
interceptor sewers served a population of 90,000.   One sewer
discharged into the ocean near Bondi, the other drained to a
sewage farm near Botany Bay.   The latter discharge was
subsequently transferred to the ocean at Malabar.
30.   On the northern side of the harbour, work commenced in
1916 on the Northern Suburbs Ocean Outfall Sewer to intercept
pre-existing harbour discharges.   This sewer extends westwards
from an outlet at North Head.
31.   Today, the Bondi, Malabar and North Head systems have
now been extended to serve some 2.3 million persons.   The
plants also serve the greater part of Sydney's industrial and
commercial development.
32.   With increase in population densities along the seaboard
and the increasing popularity of beaches, beach pollution of
sewage origin became progressively more evident.   Water
pollution control plants, providing for removal of screenings,
grit, grease, floatables and part of the settleable solids,
were commissioned in 1962 (Bondi), 1975 (Malabar) and
1984 (North Head).   Refer Figure 2.

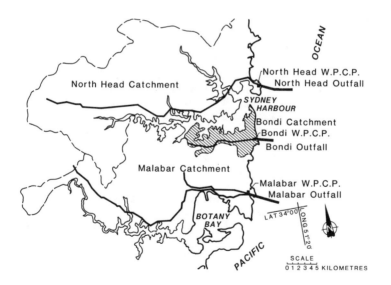

Figure 2.   Catchments and locations of Sydney's major outfalls.

33.   All three plants presently discharge to the ocean at the
nearby cliff face via submerged outlets in 10 m depth of water.
34.   In providing these facilities the Board has always
envisaged that deepwater outfalls would eventually be provided
for all three plants.

Present Cliff-Face Discharges

35. The performance of Sydney's three major outfalls is judged by the public by the condition of Sydney's surfing beaches relatively close to the outfall points. On occasions the clean beach sands may be soiled by the stranding of grease and other water-borne matter at the water's edge. At times this has caused bathers to leave the beach and contaminated sand to be removed. Analysis of samples taken from beach waters shows that bacteriological counts at times are high, confirming that beach pollution is occasionally of sewage origin. Beaches nearest the outlets experience high counts more frequently than permitted under bacteriological criteria set by the regulatory authority, the State Pollution Control Commission (SPCC).

36. Another deficiency of existing outfalls is the discolouration of ocean waters in the form of a plume extending several kilometres seaward. These plumes form on the surface of the ocean because the discharge in each case is into shallow water with very little initial dilution, generally only 10:1.

Regulatory Criteria

37. In New South Wales, criteria for ocean discharge are set down by the SPCC. These criteria are framed in terms of receiving water quality objectives and cover the following parameters:-

Physical Appearance	- no obvious visible evidence of sewage discharge by way of field, slick, grease or floatable material.
Deposition of Solids	- no accumulation of solids on the shore or the ocean floor within a distance of 1000 metres from the shore or in a depth of less than 10 metres.
pH	- pH of wastes not to exceed 8.5 or induce a variation of more than 0.1 at boundary of initial dilution zone.
BOD	- There is no specific requirements regarding BOD of wastes, however the SPCC is currently examining whether licence approvals for effluent discharge should include suspended solids limits.
Bacteriological	- in samples taken from the waters of "designated bathing areas", the geometric mean of the number of faecal coliforms is not to exceed 200 organisms/100 mL for any month between November and May. Additionally, the 90 percentile limit is set at 400 organisms/100 mL during the period November to May.
Biological	- ocean waters must be protected to retain a natural and diverse, but not necessarily unchanged, variety of marine life.

Restricted   — these include certain heavy metals, phenolic
Substances   compounds and chlorinated hydrocarbons,
                 especially pesticides and P.C.B.'s. All such
                 wastes must be controlled at their source so
                 that their concentrations at the boundary of
                 the initial dilution zone shall not exceed
                 stringent prescribed values.

38.   These criteria do not necessarily prevent the combined
discharge of digested sludge with effluent to ocean waters.

Options for Upgrading Major Outfalls

39.   Major upgrading options available at each of the three
major outfall systems were evaluated and environmental impact
statements were made available for public perusal early in
1980.   Effluent disposal alternatives were:

.   Reduced discharge by means of reuse, after further
    treatment.
.   Shoreline discharge following secondary treatment by the
    activated sludge process (with effluent chlorination).
.   Nearshore discharge following chemical precipitation.
.   Deepwater discharge of primary effluent and digested
    sludge.

40.   The statements concluded that for each of the major
outfall systems the scheme which satisfied all receiving water
criteria for the least cost comprised the discharge of primary
effluent and appropriately treated sludge through long
deepwater submarine outfalls.

41.   This conclusion was reached after comprehensive field
and office studies carried out under the direction of
Consultants Caldwell Connell Engineers.   This is reported in
the literature (ref. 3).

42.   Before proceeding to the construction stage, the Board,
in conjunction with the SPCC, invited a technical panel to
provide independent advice on the efficacy of the proposal.
The panel, which comprised Professor Poul Harremoes (Denmark)
and Professor Norman Brookes (USA), endorsed the proposal
subject to further prescribed monitoring checks.

Sludge Disposal

43.   The sludge disposal strategy needs to be viewed in the
light of the Board's overall philosophy for its three major
outfalls, namely to provide onshore treatment first and
outfalls later.   In this regard, the removal of screenings,
scum, floating grease and grit is considered vital to the
successful operation of future extended outfalls.   Reduction of
suspended solids on the other hand, is considered of lesser
importance.

44.   Accordingly full primary treatment, including sludge
digestion, was provided at Bondi and Malabar and digested
sludge was discharged to the ocean with primary effluent.
Operating experience at these plants, however, led the Board to
adopt a different strategy at North Head.   Here a "high-rate"
primary treatment plant has been built, providing for capture

of the heavier sludge fraction only (10%-20% of suspended solids) and for incineration of this sludge together with screenings, scum and grit. This option appeared to offer overall economies over full primary treatment, produced a comparable final effluent and enabled the Board to gain experience in sludge incineration on a fairly large scale. This last factor may prove invaluable should the Board find that ocean disposal of sludge is environmentally (or publicly) unacceptable in the future. Furthermore, it may have relevance to sludge disposal for the Board's numerous inland plants.

45. It is worth noting that the Board operates 38 plants which generate 115 t/d of sewage sludge (dry solids). Of this, the three ocean plants produce 75 t/d. By year 2000, daily sludge production is likely to reach 180 tonnes, virtually all of the increase being at inland plants. Studies into alternative sludge strategies has led to the Board to adopt the following sludge disposal policy:

.   Ensure, by trade waste controls and treatment, that sludge quality is suitable for utilisation on land or disposal to ocean without detriment to the environment.
.   Choose land or ocean disposal systems which give maximum economy, with preference for sludge utilisation on land where cost margins are small and land disposal is practicable.
.   Use incineration in special cases where alternative sludge disposal methods are neither feasible nor economic.
.   Continue to monitor the development of sludge treatment methods involving energy recovery.

46. For the major ocean outfalls this will mean continuing ocean disposal for the major portion of sludge production pending identification of more appropriate technology. Ocean disposal of sludge would be discontinued only if more economical alternative technology becomes available or monitoring results indicate adverse effects on the environment. Incineration would be continued at the North Head plant.

47. Signals are coming from the regulatory authority, the SPCC, that they are far from convinced that continued ocean disposal of sludge is appropriate. The Commission's draft policy is:

.   The shoreline disposal of sludge will not normally be accepted by the Commission. Subject to stringent environmental monitoring however, the Commission will permit the disposal of digested sludge with primary treated sewage effluent through extended, deepwater outfalls.
.   In all cases where sludge is disposed of to the ocean, the sewerage authority must develop and implement a chemical and biological monitoring programme to the satisfaction of the Commission. The discharge of sludge to ocean waters will be permitted to continue only if this monitoring programme establishes that there are no significant adverse environmental effects.

48. Indications are that the final policy will be less stringent and will be related to environmental needs.

49.   As noted earlier, the geographic location of the Malabar
and Bondi plants virtually prevents all opportunities for land
disposal or reuse of sludge should ocean discharge be ruled out
in the future.

Design Features of Sydney's Three Outfalls

50.   The Malabar plant is being connected to its proposed
large offshore tunnel by a horseshoe section decline tunnel on
a 1:4 grade. A change-over station for assembly of the tunnel
boring machine, spoil removal and drainage pumping will be
provided at the bottom of the decline. Both will later be
insitu lined to flow size. The main tunnel, which will be
lined in precast concrete to 3.48m diameter, will proceed on a
1:200 rising grade under the ocean to the diffuser area some 3
km offshore. The minimum rock cover will be 45m and the
maximum water depth 80m. The rock tunnel section will reduce
in cross-section over the length of the diffuser section to
maintain self cleansing velocities and connect to 28
pre-drilled risers placed 25m apart, each riser being fitted
with an 8-ported diffuser head. Figure 3 shows relevant
details.

51.   The North Head outfall will be similar to Malabar's, of
the same cross section, but with 36 risers discharging via
6-ported diffuser heads at 21 m intervals into 60 m water
depth. This outfall tunnel turns 17 degrees at about mid
length to provide a better crossing angle through 2 igneous
dykes.

52.   The Bondi outfall will be smaller and shorter than the
other two by reason of the smaller flows. The on-shore works
are to be considerably more extensive and involve an extension
to the existing underground works together with an underground
pumping station to provide driving head for the outfall. The
pumping station will be accessed by a spiral decline on a 1:6.5
grade. From the pumping station, the outfall continues on a
1:6.5 grade and passes under the cliff through a local
partly-weathered dyke. After reaching its lowest level, the
tunnel will be extended on a 1:200 upgrade in cast in situ
concrete lined tunnel (2.25m diameter) to the diffuser area.
The effluent will be discharged though 26 risers at 20m
intervals with 4-ported diffuser heads into approximately 60m
of water.

53.   The decision to adopt tunnelled outfalls with
down-drilled diffuser risers was made after very extensive
investigations (ref. 4). Drilling risers from a dry tunnel was
rejected because of greater estimated cost, higher risk and
longer overall construction time. The diffuser configuration
ultimately adopted, though requiring more down-drilled diffuser
risers than earlier proposals, proved economic and provided
much less risk against damage by anchor chains of large vessels
which often have to stand-to off Sydney Harbour. The risers
are topped by circular multi-ported diffuser heads with
removable tapered nozzles, protected by a heavy steel caisson.
Figure 4 shows relevant details.

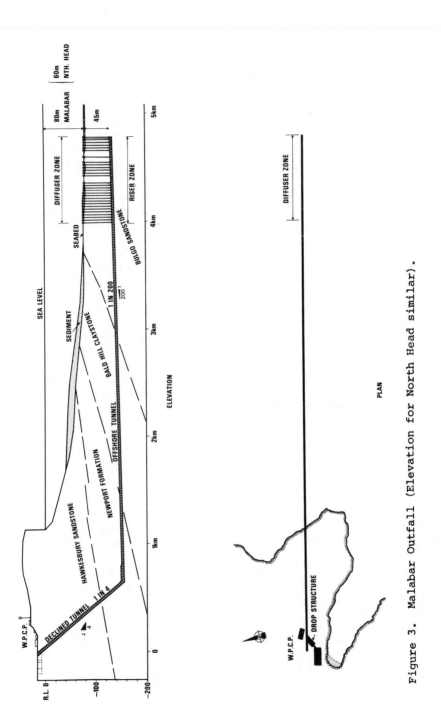

Figure 3. Malabar Outfall (Elevation for North Head similar).

PLAN

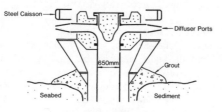

ELEVATION

Figure 4. Typical 8-port diffuser head.

54. One of the most critical operations is the connection of the tunnel to the pre-drilled riser. The risers will be installed, ahead of tunnelling operations, in a target position 12m from the centreline of the tunnel. The riser will be positively located by probe drilling from the tunnel. A 2m spur tunnel will then be excavated from the tunnel to expose the lower portion of the riser, which will be joined by pipe to the main tunnel. The spur tunnel will then be backfilled with concrete and grouted. As a safeguard against subsequent damage to the diffuser head on the seabed, and consequent (catastrophic) seawater entry into the tunnel during the remainder of the construction process, an inflated plug will be placed in the tunnel end of the pipe, the riser filled with water and pressurised to ocean pressure. The plug will remain until the tunnel is ready for commissioning.

Construction Progress

55. Work is currently proceeding on all 3 sites, the intended commissioning dates being mid 1990 for Malabar and 1991 for North Head and Bondi. All three decline tunnels are currently under construction or completed. Contracts have been awarded for construction of offshore tunnels from Malabar and North Head and for the drilling of diffuser risers. The Bondi tunnel contract will be awarded next year. Construction activities are being overlapped where practicable to achieve earliest completion. Total cost of the 3 outfalls, including onshore work at Bondi, is $300M.

PUBLIC ATTITUDES AND INTERIM MEASURES

56. In the past three years the Board has closed a gap which had been developing with its customers. This has been done by decentralising many business and operating functions, involving the establishment of a number of business and regional offices.

57. Despite this and a very active campaign to inform the public of progress with the Board's beach pollution control strategy, a recent independent survey showed that:-

. Beach pollution is very much the weak point in the Board's image.

. The majority of people sampled (55%) were unable to nominate any specific measures being implemented by the Board (11% said the Board was doing nothing and 44% were unsure).

One third did, however, have some knowledge of the outfall proposals, reflecting the impact of TV advertising.

58. Beach pollution is a matter of political concern as many coastal electorates involved are swinging seats. Publicity and progress with the outfalls, however, is not sufficient to quell public discomfit when weather conditions lead to regular incidents of beach pollution.

59. Two complementary programmes have been initiated as interim measures pending completion of the outfalls:

. The establishment of an on-the-spot advisory service to beachgoers.

. The implementation of a grease control strategy.

60. As a service to the public, to provide accurate up-to-the-minute information to the beachgoer on all aspects of the beach environment and to answer questions on what the Board is doing to combat beach pollution, a "Surfline" team has been established.

61. The team comprises 5 young staff members who are regular surf craft users with a personal commitment to eliminating beach pollution. They operate intensively during the 4-5 month bathing season, inspecting metropolitan surf beaches from early morning.

62. On occasions when there is a perceived health risk, the team alerts local health officers who may elect to close the beach in the interest of public health.

63. As grease is a significant contributor to present beach pollution problems and, later, in order to assure that the operating efficacy of the proposed outfalls is not hindered by grease levels in sewage effluent, the Board is currently implementing a grease control strategy. This entails:

. improving the capture of grease and oil at coastal treatment plants.

. reducing, at source, the quantity of grease and oil being discharged to sewers.

64. This is being achieved by simple upgrading of treatment plants, imposing more stringent trade waste discharge limits, targeting large grease dischargers and negotiating reductions, and educating the public how they can help, e.g in the kitchen.

CONCLUSION

65. Australia's excellent climate and beaches lead to
widespread recreational use of ocean waters adjoining towns and
cities. It is not surprising therefore that sewerage
authorities receive a great deal of public criticism when their
sewage disposal arrangements cause marine and beach pollution.
Long standing problems of this type are now being progressively
eliminated. Comprehensive biological and oceanographic studies
have been undertaken to enable appropriate solutions to be
found and provide data on impacts of existing discharges.
Strategies being implemented, in making use of the capacity of
the ocean to assimilate wastes, recognise the differing
characteristics of each site.

REFERENCES

1. HINWOOD J. Waste Disposal in Offshore Waters. Engineers
   Australia, May 31, 1985.
2. COX B.G. and KELSALL K.J. Construction of Cape Peron Ocean
   Outlet, Perth W.A. Proceedings Institution of Civil
   Engineers, April 1986.
3. BROWNE J.H. and WALLIS I. Investigation and Preliminary
   Design of Deepwater Submarine Outfalls off Sydney.
   Institution of Engineers Australia, Engineering Conference
   Canberra, 1981.
4. CLANCY K.G. and CARROLL D.J. Key Issues in Planning
   Submarine Outfalls for Sydney, Australia. Marine Disposal
   Seminar following 13th Biennial International Conference of
   IAWPRC, Rio de Janeiro, 1986.

7. Marine treatment, disposal and performance of sewage outfall — UK practice and expenditure

R. HUNTINGTON, BSc, FICE, FIWES, FIPHE, Head of Engineering and Operations, Wessex Water Authority, Bristol, and
P. B. RUMSEY, MSc, MICE, Manager, Sewerage, WRc Engineering, Swindon

SYNOPSIS. This paper describes the nature of the sewage outfall stock with particular reference to historical trends and discusses the scale and importance of the marine treatment option for sewage in the UK. The value of marine treatment is quantified in the context of industry capital and operating expenditure. Attention is drawn to the need to ensure that modern outfalls are designed to safeguard public health and the environment, and not purely to comply with legislation. The desirability of reviewing existing legislation to allow the development of rational criteria for design is discussed, and the issue of storm sewage overflows highlighted.

INTRODUCTION

1. The UK water industry is heavily dependent upon the use of the marine environment for the treatment of waste. Most of the old sewage outfalls around the UK coast were designed and constructed without the present day emphasis on environmental protection. Many are far too short and result in unsatisfactory conditions on local beaches.

2. The cost advantage to the UK water industry of treatment of sewage and sludge at sea is considerable, and if the industry wishes to continue to benefit from this outlet it must continue to adopt a responsible approach with regard to the environment. This requires the industry to be at the scientific and engineering forefront in developing current practice to provide the required level of protection.

3. Much of the data presented in this paper is drawn from as yet unpublished work carried out at WRc Engineering as part of the industry's wide-ranging outfall research programme.

POPULATION AND OUTFALLS CHARACTERISTICS

4. A significant proportion of the UK population resides in coastal districts (see Table 1). Historically therefore, the sea has been the most obvious route for receiving sewage effluent.

Table 1. Populations connected to outfalls and
primary treatment works (PTWs) discharging to the
sea or tidal estuaries.

	POPULATION (millions)				
		Connected to outfalls and PTWs			
	Resident	Outfalls		PTWs	
AREA	Population connected to sewerage		as % of resident population connected to sewerage		as % of resident population connected to sewerage
England	44.3	6.7	15	4.6	10
Wales	2.9	1.3	44	0.2	8
Scotland	4.9	1.6	33	1.1	23
Northern Ireland	1.3	-	-		
UK (exc.NI)	52.1	9.6	18	5.9	12

Note: Data for Northern Ireland not available.

5.   There are about 1000 outfalls around the UK which
discharge sewage to coastal waters or estuaries, either as
raw sewage, or with preliminary or primary treatment only.
6.   A useful sub-division of these outfalls can be drawn to
allow consideration of major discharges. In this context the
following definitions have been given to "major discharges".

(a) connected population/equivalent population   7000.
(b) outfall length   500 m below MLWST.
(c) discharge important for environmental reasons.
(d) significant capital expenditure committed or planned.

7.   Emergency and storm water overflows are excluded, as
are discharges giving full or partial secondary treatment.
8.   Of the 1000 or so UK outfalls, just over 500 can be
classified as 'major discharges'. These discharges, however,
serve over 95% of the total 15.5 million population connected
to outfalls. The remaining 500 discharges serve a population
of only 0.6 million.
9.   About 25% of all outfalls currently in use were
constructed prior to 1926 and some 50% prior to 1956.
10. The average length of outfalls constructed since 1960
has increased dramatically; a clear indication of the
increasing trend towards design for environmental
protection. (see Fig. 1). Furthermore, the great majority
of all outfalls completed after 1980 are served by headworks
providing at least screening of the effluent, compared to

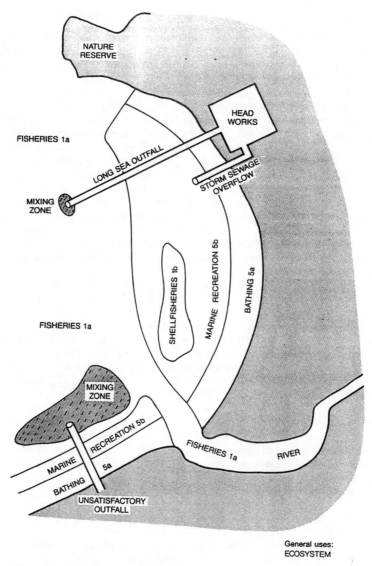

Fig. 1. Use area chart

only 20% of those constructed before 1926. There is only one record of an outfall built prior to 1965 which is equipped with a diffuser but diffuser systems are incorporated in about 50% of post-1980 outfalls.

MODERN OUTFALLS

11. For the purpose of this paper, modern outfalls are defined as those which were built or enhanced after 1970. The following discussion is based on an analysis of 88 such outfalls, which possess many distinctive characteristics.

Length below mean low water spring tides (MLWST):

12. The modern long sea outfall is designed to take full advantage of the abundant supply of dissolved oxygen and the highly dispersive nature of coastal waters, just as land treatment works use the natural purifying processes of rivers to improve effluent quality.

13. Modern sea outfalls are generally much longer than their older counterparts. The greater length ensures that they reach the deeper waters and currents necessary to fully utilize natural marine treatment processes. In combination with diffusers this results in much higher initial dilution and better secondary dispersion being achieved than was obtained by earlier designs.

14. The average length of modern outfalls below mean low water at spring tides is 540 m which is almost seven times greater than the 80 m average of outfalls completed before 1970. Nevertheless, a significant number of post 1970 outfalls (35%) terminate at or above MLWST; these shorter outfalls serve relatively small populations (average 5000). The average population served by outfalls which terminate more than 1000 m below MLWST (25% of all post 1970 outfalls) is in excess of 60,000 per outfall.

Headworks

15. About 80% of post 1970 outfalls are served by headworks providing at least screenings of the effluent. Some 70% are provided, in addition, with either maceration or comminution of the sewage. Of the latter, the use of maceration or comminution is fairly evenly split.

16. Of those outfalls equipped with macerators, about 60% return the screenings downstream of the screens; the remaining 40% return the screenings upstream.

17. Modern outfalls discharging unscreened sewage generally serve small populations; only one such outfall has been identified which serves a population greater than 2500, and this is due to be replaced in the near future. The average population served by the remaining 15 such outfalls is 1100.

Diffusers

18. A distinguishing feature of the modern outfall is the diffuser, which allows a much greater degree of initial

dilution. Diffusers are incorporated in about 40% of outfalls built since 1970.

19. All outfalls with a length below MLWST greater than 500 m are equipped with diffusers.

Materials

20. The most commonly used materials for outfall pipeline constructions are ductile iron, steel, grey iron and pre-cast concrete. Other materials used include uPVC, high density polyethelene and glass reinforced plastic. Proportions of construction materials used are shown in Table 2.

21. In certain conditions concrete-lined tunnels have been employed and these account for 5% of modern outfalls.

Table 2. Outfall pipeline materials

Material	%
ductile iron	31
steel	21
grey iron	17
pre-cast concrete	12
other	19

22. The choice of outfall material is a function of three variables:

(a) length of the outfall below MLWST - outfalls which extend less than 100 m below MLWST are predominantly segmentally constructed ductile or grey iron (about 45% and 30% respectively), whereas those extending more than 1500 m are almost exclusively either steel pipelines (60%) or concrete-lined tunnels (30%).

(b) total length of outfall from headworks - of outfalls which are greater than 2500 m in total length 65% are steel, 15% ductile iron and 15% concrete lined tunnels. Of outfalls less than 100 m in total length 75% are ductile or grey iron.

(c) diameter of outfall - for outfalls over 1 m in diameter concrete lined tunnels are the most common (45%). Steel and ductile iron (40% and 10% respectively) are also used.

WATER INDUSTRY EXPENDITURE

23. The water industry has a huge investment in its existing outfall stock. The following section presents past and future expenditure figures for both the construction and operation of outfalls and includes figures which quantify the impact if marine treatment were replaced by conventional land based treatment.

24. A distinction has already been drawn between 'major discharges' and those outfalls either serving small populations, having short lengths or involving relatively small-scale capital expenditure. As stated earlier, the 'major discharges' serve over 95% of the total population connected to primary treatment works and outfalls. There are 511 major discharges - 166 in Scotland, 280 in England and 65 in Wales.

25. The cost information presented in the following sections refers to these 'major discharges' and uses mid-1985 prices.

Capital Expenditure

26. Actual and predicted capital expenditure for the 20 year period 1975-1995 is shown in Table 3 and the regional division of total expenditure is in Table 4.

27. The results presented include expenditure on the replacement of existing assets, maintenance or improvement of existing levels of service, and for growth of development in each of the categories shown in Table 4. In detail, the types of works included are new units of all types at primary treatment works; installation of screens, maceration and other treatment units in new or enhanced existing headworks; capital works for odour or septicity control; new pumps to replace existing at pumping stations; new complete pumping stations; to intercept unsatisfactory crude sewage discharges, or to divert sewage from unsatisfactory outfalls or PTW's; repair of outfalls; addition of diffusers; lengthening of outfalls; new outfalls to replace existing overloaded or dilapidated outfalls.

28. The expenditure shown in Tables 3 and 4 has been derived from capital schemes (past or planned) associated with any of the 1000 or so UK discharges to sea (about 250 schemes in all). However, it should be noted that expenditure on major schemes at 6 locations contributes about 60% and 35% to pre-1985 and post-1985 expenditure respectively. In addition, planned expenditure includes only for schemes already planned and due to be completed in or before 1995. It does not include monies attributed to potential schemes which, if confirmed, would represent an additional cost in excess of £56 m in total over the 10 year period.

29. It is illuminating to consider the figures shown in Tables 1 and 4 in the context of total industry expenditure. Currently planned UK (exc NI) expenditure over the next 10 years is £82.6 m per annum, which is 19% of the £441.3 m per annum actually spent on all sewerage and sewage treatment in the year 1984/5 and 9% of the total of £896.8 m per annum spent on all water functions. For Britain the figures would be:-

(a) England: £51.9 m pa currently planned, which is 14% of £371.8 m pa on sewerage and sewage treatment, and 7% of the total of £756.5 m pa on all water services.

(b) Wales: £7.4 m pa currently planned, which is 59% of £12.5 m pa on sewerage and sewage treatment, and 18% of the total of £41.3 m pa on all water services.

(c) Scotland: £23.3 m pa currently planned, which is 41% of £57 m pa on sewerage and sewage treatment, and 24% of the total £99.0 m pa on all water services.

Table 3. Capital expenditure for outfalls and primary treatment works by category. (£m, 1985 prices)

	1975–1985		1985–1995		total for 20 years	
	£m	(%)	£m	(%)	£m	(%)
Extension of PTW's	14.7	(2.5)	74.3	(9)		(6.5)
Enhancement of existing outfalls	20.7	(3.5)	43.7	(5)	64.4	(4.5)
New PTW's	126.6	(21)	73.7	(9)	200.3	(14)
New LSO's	154.1	(25)	194.2	(23)	348.3	(24)
Sewerage assoc'd with the above	297.4	(48)	440.6	(54)	738.0	(51)
Total UK(exc NI)	613.5		826.5		1440.0	

Table 4. Capital expenditure for outfalls and primary treatment works (PTWs) by region. (£m, 1985 prices)

	Total		Per Annum		
	1975–85	1985–95	1975–85	1985–95	Average over 20 years
England	457.4	519.3	45.74	51.93	48.83
Wales	15.8	74.0	1.58	7.40	4.49
Scotland	140.3	233.2	373.5	23.32	18.86
Total UK (exc.NI)	613.5	826.5	61.35	82.65	72.0

Operating Expenditure

30. Operating costs are defined as all costs directly associated with the operation, maintenance and repair of works. Financing costs of the actual works are excluded. Non-recurring costs are included. The costs include

101

labour and labour on costs, first time supervision, power
including heating and lighting of buildings, fuel oil,
transport vehicles, other mobile plant, rates, buildings
and grounds maintenance, outside services, chemicals, and
any other general costs attributable to the works.

31.   The figures presented in Table 5 were drawn from
information on a sample of 'major discharges' (65 out of
511). Terminal pumping (TP) is separated out, since the
operating costs for TP include for the TP stations
themselves, the rising mains, and any pumping costs at the
works inlet. In addition, separate data on terminal
pumping has only been obtainable for a smaller percentage
(39 out of 511) of the 'major discharges'. Nevertheless,
it has been possible to derive a global figure of £0.5 per
head of population for TP operating costs, and this figure
is used in deriving the total operating costs shown in
Table 5. For less significant discharges a figure of £500
pa per discharge is assumed.

Table 5. Operating costs for outfalls and primary
treatment works (PTW's) (£m per annum, 1985 prices)

	Outfalls	PTW's	Both
England exc TP	1.4	8.2	9.6
" inc TP	4.7	10.5	15.2
Wales exc TP	0.3	0.7	1.0
" inc TP	0.9	0.8	1.7
Scotland exc TP	0.6	1.8	2.4
" inc TP	1.5	2.4	3.9
Total UK (exc NI)			
exc TP	2.3	10.7	13.0
inc TP	7.1	13.7	20.7

32.   The figures in Table 5 show terminal pumping to be
a very important part of the overall pumping costs for
marine treatment. Nevertheless, the current operating
costs are a very small proportion of the total operating
costs for sewerage and sewage treatment. The total UK
operating costs of £13 m (exc TP) represents only 2% of
the annual operating costs of sewerage and sewage
treatment, and 0.9% of the annual operating costs for all
water functions.

THE VALUE OF OUTFALLS TO THE INDUSTRY
33.   It is sobering to consider the impact on industry
expenditure which would arise if the current practice of
marine treatment via outfalls and PTW's were to be
abandoned, and primary or primary and secondary treatment
of all sewerage flows became the norm.

34. The additional capital expenditure required to replace outfalls and primary treatment works can be approximately assessed by making the following assumptions:

(a) if an outfall already exists it will be used to discharge effluent which will be fully treated.

(b) if an outfall is currently planned it will be abandoned and the proposed capital expenditure diverted to the construction of primary treatment works.

(c) - land is available for construction of a treatment works at existing outfall or primary treatment works sites.

(d) additional sewerage required will amount to 50% of the cost of constructing a primary treatment works.

35. In order to enable all discharges to receive primary treatment, through works constructed over a 10 year period from 1985, it is estimated that an additional £84m pa capital expenditure would be required for the UK (excluding NI). Additional operating costs are estimated to be £21 m pa. The additional operating costs include any extra pumping due to a rationalisation of sewerage and the cost of operating new sewerage systems.

36. For the provision of primary and secondary treatment over a 10 year period, the corresponding figures are £166 m pa (capital) and £49 m pa (operating). The figure for operating costs includes for treatment and disposal of sludge in accordance with current practice.

37. The benefits derived in respect of operating costs are shown in Table 6. These figures demonstrate the value to the UK water industry of maintaining the practice of marine treatment of sewage. The increases in capital expenditure needed to respond to an embargo on marine treatment via outfalls would require either very large increases in charges to finance the works, or wholesale transfer of capital expenditure planned on other water services. Neither of these is likely to be acceptable; the only alternative would be the introduction of change over a much longer period than 10 years.

PUBLIC HEALTH, ENVIRONMENTAL PROTECTION AND THE LAW

38. Current legislation involves the following Acts and Directives:-

Control of Pollution Act
Dangerous Substances Directive (CEC)
Titanium Dioxide Directive
Mercury Directives (2)
Cadmium Directive
Hexachlorocyclohexane Directive
Bathing Waters Directive (CEC)
Shellfisheries Directive (CEC)

39. The Control of Pollution Act Part 2 (COPA II) gives the water authorities power to enforce the EEC directives listed above. Furthermore the discharge of any material

or effluent to a sewer must have the consent of the local
water authority. In the case of the water authority
wishing to initiate a discharge to sea or tidal estuary,
the approval of the Department of the Environment (DoE)
must be sought. In either case the consent must be
referred to the Ministry for Agriculture, Fisheries and
Food (MAFF) who will assess the impact of the proposed
discharge on fishery interests.

Table 6. Comparison of operating costs

| | Outfalls (inc TP) | PTWs (inc TP) | Outfalls and PTWs (inc TP) | Total if all discharges were to receive | |
				Primary Treatment	Primary and Secondary Treatment
Population (000000) connected	9.6	5.9	15.5	15.5	15.5
Total operating costs p.a. (£m)	7.12	15.64	20.8	41.8	70.2
As % of per annum operating costs on all sewerage and sewage treatment	1.2	2.2	3.4	6.9	11.5
Total operating costs per capita per annum (£m)	0.76	2.29	1.33	2.68	4.51

40. A frequent and justifiable criticism of the EEC
Directives is that they are both inconsistent and
difficult to apply to design. The Directives express
limits on the concentrations of various pollutants in
terms of annual average concentrations, 95, 90, 80 and 75
percentile concentrations, monthly averages, monthly
maximum loads and maximum concentrations. Furthermore,
the Directives are framed to facilitate monitoring rather
than design and, as a consequence, place unnecessary
burdens on the designer. For example, in order to comply
with the Bathing Waters Directive the designer is obliged
to assess at least four different sets of conditions; this
is over-onerous and does not afford any greater degree of

environmental protection as less resources are available
for the in-depth analysis of any one case.

41. Criteria developed for ease of monitoring are
inappropriate for outfall design. A unified and
design-orientated approach is required in setting the
limit concentrations of pollutants. This is particularly
important if unnecessary and time-consuming calculations
are to be avoided.

OUTFALL DESIGN

42. The success of many modern long sea outfalls bears
witness to the environmental performance of marine
treatment via properly designed and constructed outfalls.
However, despite the very considerable UK experience of
sea outfall design there remain areas of uncertainty which
are generally dealt with by adopting a conservative
approach and good judgement.

43. The area of greatest doubt is the determination of
the level of protection required to safeguard the
environment. The situation is not made easier by the
uncertainty, discussed above, in the application of EEC
Directives to design.

44. The UK's recent decision to set Environmental
Quality Standards (EQO's) to limit the concentrations of
certain contaminants in the receiving water is a step is
the right direction. Nevertheless, a framework is
required which will allow a rigorous and objective
approach to outfall design.

45. Current practice dictates that the controlling
authority (i.e. water authority or DoE) designates
specific areas of water as being required to comply with
particular Directives, or with specified Environmental
Quality Objectives (EQO's).

46. A more rigorous approach, recently set out in the
WRc publication "Outfall Design Guide for Environmental
Protection" involves the clear definition of the uses of
receiving waters in the vicinity of the outfall. EQO's
are identified as applying to each broad Use category,
together with the appropriate EQO's. The identified Uses
are set out in Table 7, together with their associated
EQO's and the EQOs are described in Table 8.

47. Use of these concepts is illustrated in Fig. 1.
Further details are contained in the "Outfall Design Guide
for Environmental Protection". A similar framework is
currently under consideration by the Water Authorities
Association for estuaries.

48. Inherent in this approach is the need to allow a
reasonable mixing zone for a discharge to mix with the
receiving water. The "mixing zone" is the area around a
discharge point wherein the EQOs of any other Use areas
may be exceeded, and some level of environmental damage
may occur. The decision on whether or not a "mixing zone"
is reasonable in size is a matter of judgement for the

controlling authority who should be involved in discussion
from an early date.

Table 7. Use areas and their respective EQOs

Use Area	EQO
Fisheries	1a, 2
Shellfisheries	1b, 2
Ecosystem	2, 3
Abstraction	4
Bathing	5a
Marine Recreation	5b
Mixing Zone	6

Table 8. Environmental quality objectives

EQO NO.	EQO
1	Human food source protection (a) fisheries (b) shellfisheries
2	Fish and Shellfish protection
3	Marine animal protection
4	Industrial abstraction protection
5	Bathing and contact water sports protection (a) bathing (b) contact water sports
6	Public nuisance prevention – Aesthetic considerations.

STORMWATER OVERFLOWS

49. In the UK the majority of sewerage systems collect
both domestic sewage and storm water runoff. It is not
economic to design either an outfall or its headworks to
cope with storm water flow and, in any case, oversizing of
an outfall or headworks will lead to operational problems.

50. It is normal practice to divert flows greater than
Formula A to a storm water overflow. These overflow
limits are largely arbitrary and are not functions of the
polluting effect of the discharge on the environment.
There seems little point, however, in investing large sums
of money in designing and constructing outfalls to comply
with EEC standards if frequent overflows will jeopardise
that compliance.

51. In order to fully assess the effects of storm water
discharges on the environment the designer should use
predictive dispersion and storm flow models. Having done
so the maximum overflow discharge rate and length of
overflow pipe consistent with the EQOs may be determined.

The impact of the overflow on the environment may then, if necessary, be reduced by using one or more of the following:

(a) treating the storm flow to remove solids
(b) providing storm sedimentation tanks as at land treatment works.
(c) providing storm water storage at say the 65 1/head recommended by the technical committee on storm overflows (HMSO 1970).
(d) storing the first flush for subsequent treatment.

52. In order to comply with aesthetic standards storm overflow should at the very least be screened.

CONCLUSION
53. There are about 1000 sewage outfalls around the UK coast, serving some 30% of the population. The nature and performance of new outfalls built in recent years has changed dramatically for the better through better location, increased length, construction of diffusers and the introduction of improved headworks treatment facilities.
54. Water industry future capital expenditure on outfalls and PTW's has been identified as being £83m pa over the next 10 years, with operating costs of £21m pa (including terminal pumping). The projected capital expenditure represents about 19% and 9% respectively of the planned capital expenditure on all sewerage and sewage treatment, and on all water services. The low figure for operating costs (about 2% of all operating expenditure on sewerage and sewage treatment) reflects the efficiency from a financial viewpoint of marine treatment via outfalls.
55. It is estimated that the capital cost of providing primary treatment for all discharges to coastal waters, would be about twice the currently projected expenditure; to provide primary and secondary treatment the projected capital expenditure would need to be tripled.
56. Current legislation could be much improved to allow more effective outfall design, in particular with respect to consistent limit concentrations of pollutants. A more rigorous and logical framework for design can also be achieved by the introduction of procedures which clearly define, at the initiation of design, the Uses required of all waters in the vicinity of the outfall.
57. Particular attentions needs to be paid to the design and operation of storm sewage overflows.
58. The UK Water Industry is increasingly providing outfalls which afford a greater degree of environmental protection. This is evident in the trend of constructing longer outfalls equipped with diffusers which take full advantage of natural marine treatment processes, and by

the fact that older outfalls are being replaced at an accelerating rate.

59. The water industry has responded to the public concern for the environment. The full extent of this response will not be apparent for some years because of the time required to design and construct better designed sea outfalls. However, it can be confidently expected that marine treatment schemes will be accepted as an environmentally safe solution to sewage disposal.

ACKNOWLEDGEMENTS

The data on water industry expenditure presented in this paper is drawn from studies commissioned by WRc Engineering and carried out by James C Coleman, Consultant.

REFERENCES
1. NEVILLE-JONES P J D and RUMSEY P B. The Engineering Design of Outfalls for Environmental Protection. IPHE Symposium - The Construction and Maintenance of Long Sea Outfalls. Portsmouth, 1986.
2. Outfall Design Guide for Environmental Protection. WRc Report ER September 1986.
3. Technical Committee on storm overflows. HMSO 1970.

Discussion on Papers 5-7

MR HUNTINGTON and MR RUMSEY, Paper 7
Raw sewage is extremely variable in its composition but, provided the discharge of non-domestic sewage to sewers is properly controlled, sewage should contain few contaminants which are in sufficient strength to be troublesome.

For land-based treatment, where the effluent is discharged into a river or stream, biological oxygen demand, suspended solids and ammonia, are usually critical, while for marine treatment, bacteria levels are important. Land-based treatment only reduces bacteria levels by one order of magnitude (Fig. 1) and a beach close to the mouth of a river can be seriously affected by high bacteria levels carried by that river. The combined effect of bacterial die-off during treatment and dilution on receipt in the river generally ranges between only 80 and 160, which is very small compared with dilutions of 5000 plus which are achieved with properly designed marine outfalls.

The potential for dilution and bacterial decay in UK coastal waters is immense. Partial or full treatment of sewage prior to discharge has little impact on bacterial concentrations during secondary dispersion in the sea (Fig. 2).

A properly integrated design should involve examination of the properties of the receiving water, its use, the sewerage system discharging to the outfall and the treatment processes at the headworks. Attenuation in the sewerage system can now be properly evaluated using verified computer models, often leading to more cost-effective or environmentally acceptable schemes. Environmental standards can only be achieved if the headworks provide for removal of grit and solids by fine screens.

With a properly designed outfall, four hours of secondary dispersion easily achieves the 5000 dilution required to meet the EEC bathing water standard, irrespective of the initial dilution (Figs 3 and 4).

The time taken for 90% of the bacteria to die (T_{90} value) should be chosen to take account of local conditions and the fact that the decay of some species of bacteria is well in excess of that for total coliforms.

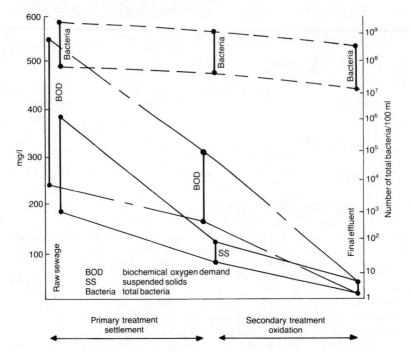

Fig. 1. Effectiveness of land-based treatment (after White, 1980)

The design process will be aimed towards ensuring that

(a) the outfall is located correctly
(b) the diffuser design is right
(c) the overall design will meet short-term and long-term
     operational needs and standards
(d) there is flexibility to help meet unforeseen future
     requirements.

Sewage and storm outfalls need to be considered together,
particularly bearing in mind that the first flush during
storms can carry a very high polluting load.

Storage may be used with advantage to restrict discharges
through a main outfall over a short length of the tidal
cycle where adverse conditions might otherwise result.

Monitoring the performance of marine treatment works,
together with receiving waters, is essential to ensure the
well-being of the processes involved, with particular
attention paid to identifying any adverse trends which might
be developing. The introduction of flow and pressure
recording facilities at the headworks, and on selected
risers along diffuser sections can provide valuable
information on the performance of the systems. For
monitoring the diffuser section, a data logging package has

been developed which can be installed by divers and allows
readings of velocity and salinity to be taken at close
intervals over a period of up to three or four days.

Three outfalls have now been monitored in this way and the
results are being used to calibrate numerical models of
outfall hydraulics. This will allow more confident
predictions of outfall performance, and the development of
better control over the treatment of effluent at sea.

Marine treatment, properly designed, constructed and
operated, is effective and can play a full part towards

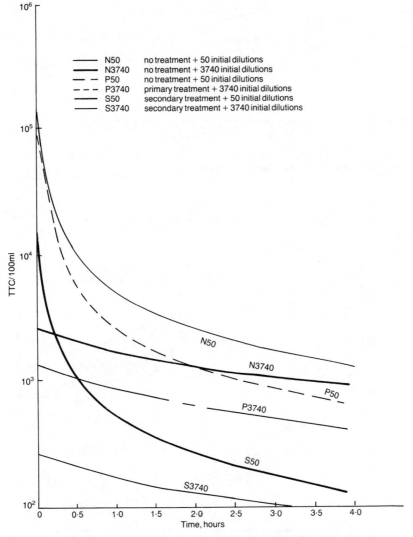

N50	no treatment + 50 initial dilutions
N3740	no treatment + 3740 initial dilutions
P50	no treatment + 50 initial dilutions
P3740	primary treatment + 3740 initial dilutions
S50	secondary treatment + 50 initial dilutions
S3740	secondary treatment + 3740 initial dilutions

Fig. 2. Effects of different combinations of treatment on
bacterial concentrations in secondary dispersions

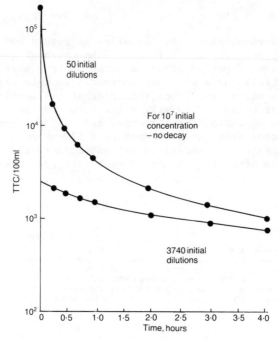

Fig. 3. Concentration of bacteria during secondary dispersion with no decay (after Lee, 1987)

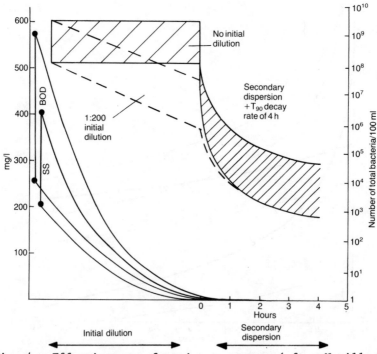

Fig. 4. Effectiveness of marine treatment (after Neville-Jones, 1986)

ensuring that sewage is returned to the environment in an
acceptable manner at a cost people are prepared to pay

MR M. KING, Land & Marine Engineering Ltd
Paper 7 relates primarily to post 1980 outfalls in the UK.
But we have, of course, been designing and building marine
outfalls in Britain for nearly 30 years, and the basics for
the construction of these long sea pipelines have not
changed radically over this period. The Paper is based on
well established practice and on work by WRc which, although
not officially published, has been discussed at earlier
symposia. I do not say this in any way critically - but it
is interesting that references in Paper 7 are limited to
three, suggesting, perhaps, that this is an original work.
    It is always good to measure or quantify the performance
of existing outfalls. It is this particular role that Mr
Huntington and Mr Rumsey have played in, I think, a masterly
way. They have taken stock and pointed to design and
operational areas in which refinements could usefully be
made. In conclusion, they estimate that to provide primary
treatment for all discharges to coastal waters - I take it
that they are referring to future discharges - it would cost
twice the currently projected expenditure. The effect of
this, it seems to me, would be to cut by half the number of
outfalls planned in current budgets - assuming the normal
practice in the UK of providing pre-treatment by screening
and some form of maceration.
    To provide full treatment prior to sea discharge would
triple the cost, or reduce budgeted outfall construction to
about one third of the number planned. I understand that it
is maybe even near the point of regulation that sewage be
given full treatment before discharge to sea. If this is
true, then it is totally unrealistic in the light of
operational experience with long sea outfalls for domestic
sewage built since 1969. I believe that those who are
advocating full treatment followed by marine treatment are
casting reality aside and possibly retarding outfall
construction programmes by up to 30 years, unless, which I
doubt, our European friends have found a mountain of money
which can happily be thrown into the sea!
    I hope that it is not too late to think through this major
issue yet again, and above all not to generalize in deciding
on a policy which could have far-reaching effects.

MR M. D. McKEMEY, Lewis & Duvivier
In Table 2 of Paper 6, the entry for Newcastle, Burwood
Beach is incomplete. The proposed construction effluent
outfall is in fact a 2 km long submarine tunnel, some 2.5 m
in diameter driven 60-80 m below sea level with upward
drilled diffuser risers topped with eight port heads in some
20 m depth of water.
    The construction is presently well advanced with the
tunnel approaching the diffuser positions. The lowest

tender received for the (vertical 90 m deep) shaft, tunnel
and diffuser was some $11 million. The estimated total cost
for secondary treatment works plus outfall was of the order
of $30 million.

Under the legislation of the various Australian states a
formal environmental impact investigation is obligatory for
most significant development projects including sewage
outfalls and works. Would the Authors comment on the value
of the wide-ranging and comprehensive environmental impact
investigation, which is a significant aspect of Australian
practice?

It has been mentioned that problems result from effluent
plumes being trapped below the sea surface due to
densimetric stratification. It is, however, intended in the
Sydney outfalls to exploit this phenomenon for aesthetic
reasons during the summer months. Could the Authors
comment?

MR P. NEVILLE-JONES, Binnie and Partners
I am intrigued by the inclined tunnels on the Sydney
outfalls. These tunnels must have drainage difficulties
during construction and once built will have a potential
hydraulic problem. On pump start-up or following any rapid
increase in flow, the waste water will start to oscillate.
For the Malabar outfall a period of around four minutes
might be expected. These oscillations are not revealed by
physical models, but would be by conventional surge
analysis. These oscillations result in flow reversals and
might lead to other problems.

I would therefore like to ask what provision is made for
drainage during construction and to what extent have surges
been considered.

MR J. A. WAKEFIELD, Coastal Anti-Pollution League
Twenty years ago the CAPL had an enquiry from an
environmental group asking how to stop the construction of a
crude sewage outfall over the Bondi Beach. We advocated the
publication of a Golden List. We subsequently heard that as
the bulldozers dug the trench for the outfall where it
crossed the beach, so the pressure group hired a bulldozer
to fill it in. It seems that the environmental group were
justified in their action. What kind of opposition is there
these days?

DR J. A. CHARLTON, Consultant
As a specialist consultant to the Sydney Metropolitan Water
Sewerage and Drainage Board, I recall on my last visit to
Sydney becoming involved with the problem of grease
accumulation in the outfalls, and the resultant grease balls
which could form after stripping from the tunnel soffit.
Apparently, the grease balls had the potential to be bigger
than the outfall diameter, and we had an amusing morning re-
designing the seaward riser-port combinations to accommodate

items this large. Does the Author of raper 6 regard this as
a potential problem generally in outfalls, and will primary
treatment minimize the problem?

MR P. N. PAUL, John Taylor & Sons
Table 2 in Paper 5 gives details of six outfalls in France.
I appreciate that most of them discharge into relatively
deep water but, with one exception, all of them are
relatively short. They are certainly much shorter than many
UK outfalls. Can Dr Quetin tell us what form of treatment
is given to the sewage before discharge, particularly on
those outfalls in the Mediterranean, and can he also give
some idea of the performance of these outfalls? I would
also be interested in any details that the Author could give
on the diffuser systems used on these outfalls.

I was interested to hear Mr Whyte's reference to grease
balls in the Sydney outfalls. While the connection of the
riser into the lower half of the outfall may prevent
problems with grease balls in the short term, what
confidence does he have that grease balls will not
accumulate in the upper half of the outfall and then get
larger and larger before finally being drawn into the
risers?

In addition, could Mr Whyte explain what specific measures
have been taken either to prevent saline intrusion or to
ensure that sea water intrusion can be purged from the
outfalls?

MR D. W. HARRINGTON, WRc Processes
Could Mr Whyte expand on the problem associated with grease
and the effect of the material on the performance of the
headworks equipment?

What type of preliminary treatment plant is installed,
particularly the screening equipment (size/mesh), and what
do you do with the screenings? I would like to ask Mr Whyte
and Dr Quetin, if primary and secondary plant is installed
to pre-treat the sewage, if they take into account the
reduction in initial bacterial count and design shorter
outfalls?

Generally, have the Authors considered the effect of storm
waters and what are the design considerations? WRc are
particularly concerned with inadequate screening of storm
waters and the problem of visual pollution caused by
persistent plastics.

MR W. G. G. SNOOK, The Kenny-Snook Association
Mr Whyte referred to the formation of grease balls in
tunnels and the obvious problems caused by this basic
malfunction. Insofar as piped outfalls are concerned, it is
my experience that the formation of grease or slime (as a
result of discharging disintegrated raw sewage, screened
sewage and sometimes even following primary settlement) can
be prevented by ensuring at design stage that velocities

during discharge exceed settling out velocities.

In the case of intermittent discharge or periodically low flows, the normal flow should provide velocities in excess of that required to re-agitate settled solids. In this respect I have always attempted to ensure that velocities should exceed 1.5 m/s in the main bore and 2 m/s in the risers and ports.

The large majority of submarine pipelines and diffuser systems I have designed rely on intermittent discharge to provide constant diffusion characteristics (ref. 1). Regular inspection of main bores and diffuser systems (with decreasing main bores designed on this principle) have shown no build-up of grease of slime whatsoever.

In tunnels, constructional economics very often dictate much larger diameters than those required to accommodate the ultimate maximum design flow at self-cleansing and re-agitating velocities.

There is also a general, and in my opinion misguided, obsession with the attainment of very low port velocities in an attempt to give Froude numbers as near unity as possible. Not only does this practice tend to encourage the formation of grease or slime, but also it induces settlement of solids and intrusion by sea water and debris contained therein. This, in turn, causes progressive blockage of the diffuser system from the seaward end.

Settlement in both the diffuser and main bore will continue until the system eventually balances itself hydraulically: until the number of discharging ports and the diameter of the main bore reduce sufficiently to provide the necessary self-cleansing and re-agitating velocities. Inherent in this final hydraulic balancing will be a dramatic reduction in operating efficiency in that the initial dilution factors achieved will be greatly reduced compared with those originally intended and certainly much less than that of a properly designed diffuser system.

In the case of tunnels I believe there is a strong case for not only laying pipes within the tunnel (including a reducing bore on the diffuser section) with diameters such as to induce self-cleansing and re-agitating velocities, but also to provide intermittent high velocity discharge at times of low inflow into the headworks (ref. 1).

I have always considered that, to comply with the EEC bathing water standards, the effluent from a secondary treatment works would, in the majority of cases, have to be discharged to sea through a submarine pipeline and diffuser system. With reference to Paper 7, does the cost of secondary treatment include the cost of a properly designed submarine pipeline and diffuser system?

Mr Rumsay described the monitoring device used to collect data for future mathematical modelling and appraisal. Such a device has been installed on the Eastbourne outfall which I believe has a constant diameter of the main bore in the diffuser section and is therefore a prime candidate for blockage as a result of the process I have described. I would therefore have thought it dangerous to include data collected from such an outfall in the production of a mathematical model for general application for design purposes.

I would like to appeal to the steering committee at the WRc to reinvestigate and enhance the already excellent work carried out by WRc, HRS and others (refs 2-7) relative to the formation of grease and slime in pipelines and, in particular, research into the velocities (relative to diameter and roughness of pressure pipes) necessary to prevent grease and slime formation and to re-agitate settled solids.

MR G. F. GREAVES, Ozotech Ltd
Having engineered schemes to protect the quality of bathing waters, fisheries and the marine environment, did the Authors give due consideration to protecting the likely high amenity value of the areas adjacent to the landward end of the schemes, especially with regard to the potential of noise and odour from the headworks?

Furthermore, the possibility of disinfecting effluents discharged to sea was mentioned and I am aware that Mr Huntington had, at one stage, considered disinfection at Holdenhurst STW. What criteria would the Authors therefore apply in assessing the need for disinfection and what is their view regarding the discharge of persistent disinfectants and their reaction products to the marine environment?

DR P. J. MATTHEWS, Anglian Water
Options which should be taken into account in a feasibility study include middle ground involving partial treatment - not necessarily primary sedimentation. I have in mind chlorination at Weymouth, the activated line - Clariflow process at Sandown on the Isle of Wight, and recent studies on UV graduation. What are the Authors' experiences and views of these options?

MR F. N. MIDMER, Director of Technical Services, Southern Water Authority
There is a considerable risk that, in providing storm overflows to avoid town centre planning, beaches will not fully comply with EEC standards. In the UK we have some reassurance in the fact that, when it rains, most of our beaches are not being used for bathing, but in France and Australia this might not be true. I would be interested to know how storm water is dealt with, bearing in mind that a short outfall is liable to give rise to relatively high bacterial levels on the foreshore. I feel that more thought is going to have to be given to this aspect of the matter to avoid future problems.

DR QUETIN, Paper 5

In answer to Mr Midmer's question, beaches and bathing spots in France are seldom used when it is raining. It has indeed been noted that rain brings a momentary pollution flux through rivers and estuaries. However, the situation along the Mediterranean coast is different from that in Brittany.

In the first case, the rivers are subject to severe drought in the summer, and the water stagnates and deteriorates in pools along the river beds. Summer storms are violent and the flood discharges considerable, with a strong flushing effect.

In Brittany, the bacterial pollution brought by surface run-off is the result of agricultural practice. Intensive stockbreeding in the region leads to the spreading of excessive quantities of liquid manure, which nature cannot recycle quickly enough. Rainfall causes leaching of this liquid manure and pollution of streams and rivers in the region. Strict legislation is envisaged to control this type of pollution.

It has also been noted in France that overcrowding of beaches gives rise to direct bacterial pollution by the bathers themselves. This pollution is measurable not only in the water but also on the sand itself.

Finally, another question clearly arises. Up to what distance from the beaches must the bacterial pollution standards be respected? Some bathers swim out to a considerable distance from the beach, while windsurfers can go even further offshore.

MR WHYTE, Paper 6

My thanks go to Mr McKemey for bringing us up to date on the progress with the Burwood Beach Outfall in Newcastle, New South Wales. Public pressure has very much influenced the outcome in this case.

Regarding the value of environmental impact statements I believe they provide a useful public interface for major projects, often leading to modifications in the proposal in the public interest which can result in gaining public support at little additional cost. The value can be lost, however, if there is a long delay before the project is undertaken.

On the subject of frequency of plume submergence, the design of Sydney's outfall/diffuser configuration is largely dictated by beach faecal coliform criteria and particularly the winter requirement of a 90% value of 200 organisms/100 ml. Frequency of submergence for this configuration was as high as 96% for the bathing season (November to May) and between 10% and 35% (dependent on outfall) for the rest of the year. In other words, a strong thermal gradient is persistent for most of the year. Dr Quetin's warning about trapped plumes impairing dilution and natural self-purification is acknowledged. However, Sydney's comprehensive oceanographic studies showed that subsurface

currents and dissolved oxygen levels were consistently high.
These factors were taken into account in the design of
Sydney's outfalls.

In reply to Mr Neville-Jones and Mr Paul, regarding
potential hydraulic and saline entry problems, such problems
were recognized and extensive model work was undertaken by
the British Hydromechanics Research Association, and locally
to arrive at a satisfactory design. In particular, a great
deal of effort and expense was applied to solve the salt
water removal problem. Papers have been published on this
subject. Regarding water-hammer effects, studies showed
that the most severe case would be a negative pressure
effect at Bondi caused by sudden shutdown of all three
pumps. Safeguards included the provision of a vent-shaft
and a refluxed bypass.

On the subject of drainage of the steeply declined
tunnels, strata have been relatively watertight at all three
sites and no problems have been experienced to date using
conventional pump arrangements.

Mr Wakefield's question on action groups reminds me of a
group advocating pumping sewage inland over the Blue
Mountains where rainfall is less than on the coastal strip.
Despite the fact that costings indicated a factor of at
least ten over the ocean disposal options, this group
remained unconvinced, saying that we should pump inland as a
matter of principle, irrespective of cost. Other more
sensible pressure groups included the surf-board rider
lobby, which is inclined to accept with some reservations
our Surfline advice while anxiously waiting completion of
the outfall project, and Greenpeace who are focusing on
heavy metal concentrations in sewage effluent. Our policy
of increased source control is a useful answer to such
criticisms.

As there has never been any intention for the Bondi
outfall to cross the beach, I can only assume Mr Wakefield's
reference to problems over a trenched outfall probably
relates to a proposed storm water drain.

Dr Charlton, Mr Snook and Mr Paul express concern about
grease accumulations and later stripping leading to large
grease balls interfering with performance of diffusers.
Certainly we had many problems with huge grease balls on
beaches before implementing primary treatment and source
control. Source control has halved raw sewage grease
concentrations. Primary treatment is removing 90% of
floating grease and about 30% of total grease. Today's
residual grease is largely dispersed and unlikely to ball to
the extent of yesteryear. We have adopted, in my view, a
belt and braces approach by providing an escape route for
stripped and balled grease via the furthermost diffusers,
which are the only two running off the tunnel soffit.
Regarding velocities in the tunnels, these are of the order
of 0.7 m/s at PDWF, and the tunnel is tapered over the
diffuser length to assure self-cleansing. Tunnels cater for

storm flows of at least twice PDWF and this will assure
regular flushing. Flow velocities in vertical risers will
be of the same order of magnitude, some being capped for
initial flows to achieve higher velocities at present
loadings. Velocities in diffuser nozzles will, of course,
be substantially greater.

In reply to Mr Harrington's questions, primary treatment
is provided at all three sites. At North Head,
sedimentation time is shorter than conventional; details are
set out in the Paper. All three plants have mechanically
raked bar screens (15 mm opening) which receive all storm
flows. Screenings are incinerated. Regarding outfall
length, the Board considered options including secondary and
chemical treatment with shorter outfalls. These schemes
were rejected on cost considerations. All three outfalls
have been designed to meet beach faecal coliform criteria
without chlorination. These criteria include a geometric
mean of 200 organisms/100 ml (outfalls should achieve 10), a
90% in summer of 400 (expect to achieve better than 80) and
a 90% in winter of 2000 (barely achievable).

Mr Harrington and Mr Midmer referred to potential storm
overflow problems. The Sydney Water Board's three major
coastal systems are not combined but still receive
substantial storm flows at times. In events when incoming
flows exceed the capacity of the deepwater outfalls,
provision is made for bypassing excess flows to existing
cliff-face outfalls. At such times, there is little
activity at beaches. However, surf-board riders may seek
the larger waves often found at this time. Fortunately the
distance from cliff face outfalls to board riding areas is
considerable.

Regarding Mr Greaves' comments on potential noise and
odour problems, Malabar and Bondi plants are entirely
underground (apart from digestion facilities) and odour
control is therefore reasonably easy if ever it is required.
Furthermore, all except Bondi are well separated from nearby
development. Noise has never been a problem and odour has
not been a major issue to our neighbours.

MR HUNTINGTON and MR RUMSEY, Paper 7
In answering Mr Greaves' and Dr Matthews' request for our
views on the effects of chlorination on the environment, we
would refer to the toxicity trials carried out by WRc on
rainbow trout using chlorinated sewage. Samples of the
chlorinated effluent failed to show the presence of organo-
chlorines other than that found in the incoming sewage and
an analysis of organs taken from the trout also indicated
that organo-chlorines were not present.

In 1978-79 and 1985 Wessex Water also carried out
biological surveys in Weston Bay where chlorinated sewage is
discharged to try to detect any changes in flora and fauna
and this data was compared with that produced in 1940 by the
Bristol Naturalists Society. No diversity in the inter-

tidal species resulting from the chlorinated sewage was found. Biannual biological studies are continuing.

On the question of use of the Clariflow process as used at Sandown, from the information at present available the process can be expensive and, more importantly, to obtain a reasonably large bacteria kill, values of pH of about 11 are required and this generates problems in relation to discharging the effluent and disposal of sludge.

With regard to Mr Snook's question about whether the cost of secondary treatment includes the cost of a properly designed submarine pipeline and diffuser system, the cost comparison given in the Paper is based on the capital costs associated with upgrading existing outfalls if full secondary treatment is required at the headworks. It has been assumed that the existing outfall can be used for discharge of the treated effluent. However, this may lead to operational and/or maintenance problems to the extent that a new outfall may be necessary. No allowance has been made for design and construction costs. Similarly, where a primary treatment works is planned which will require upgrading to secondary treatment, no cost for the outfall has been included. This is because the length and location of such an outfall will be site specific and therefore cannot be readily quantified. The use of diffusers on discharges from secondary treatment works is an interesting point. We do not believe they are widely used at present but if the EEC bacterial standards are to be met they may be required. Again, no cost for a diffuser section has been included.

Referring to the question concerning the monitoring device installed on the Eastbourne outfall to collect data for future mathematical modelling and appraisal, the outfall pipeline at Eastbourne has an internal diameter of 1067 mm with no tapering section at the diffuser. There are 25 risers of which the seaward five or so appear not to operate correctly due to the ingress of salt water.

As part of their research work, WRc are developing a computer program to predict sediment movement in an outfall. The outfall at Eastbourne was chosen for the field trials because it was believed to be partially blocked. The main purpose of the trial was to record the depth of sediment in the outfall, and the flow regime that operated at the site. This data is being used to calibrate the sediment model. The initial findings suggest that the model can be used in a qualitative way but that currently it fails to predict sediment build-up quantitatively.

Referring to the question on the formation of grease and slime in pipelines/outfalls, WRc appreciate the need for further work on this but at present have no programme of research into this area. Reference 8 concerns the effect of deposits on the hydraulic roughness of sewers but not the biological build-up of grease and slime. WRc's present involvement is to keep a watching eye on the subject.

REFERENCES
1. Snook W. G. G. Submarine pipeline and diffuser design.
   In Long sea outfalls for sewage disposal. Institution
   of Public Health Engineers, Wales District Centre,
   Cardiff, 1979.
2. Gardiner I. M. The effect of slime growth on the
   hydraulic roughness of sewers. Publ. Hlth Engr, 1979,
   vol. 7, Jan.
3. Bland C. E. G. et al. Some observations on the
   accumulation of slime in drainage pipes and the effect
   of these accumulations on the resistance to flow. Publ.
   Hlth Engr, 1975, vol. 13, Jan., 21-28.
4. May R. W. P. Deposition of grit in pipes. Hydraulics
   Research Station, Wallingford, 1975, INT 139.
5. May R. W. P. Sediment transport in sewers. Hydraulics
   Research Station, Wallingford, 1982, INT 222.
6. Perkins J. A. and Gardiner I. M. The effects of slime
   growth on the roughness of sewers. Proceedings of 17th
   Conference, Paper C32. International Association for
   Hydraulics Research, Baden-Baden, 1977.
7. Perkins J. A. and Gardiner I. M. The hydraulic
   roughness of slimed sewers. Proc. Instn Civ. Engrs,
   Part 2, 1985, vol. 79, Mar., 87-104.
8. Henderson R. J. A guide to hydraulic roughness in
   sewers in 1984. Water Research Centre, Medmenham,
   Report ER131E.

8. Sewage sludge disposal at sea — options and management

M. M. PARKER, Ministry of Agriculture, Fisheries and Food, and
A. D. McINTYRE, Department of Zoology, University of Aberdeen

SYNOPSIS. Reasons for choosing the marine option for sewage
sludge disposal are reviewed and management considerations,
including site selection and monitoring, are discussed.

INTRODUCTION
   1.   Sewage is a waste which cannot, for the most part, be
reduced or eliminated by process changes at source. It has
always been a matter of concern, causing injury to health and
aesthetic offence. Modern management practice deals with
these problems first by partitioning sewage from the human
environment through sanitary engineering and sewerage, second
by various degrees of treatment to reduce deleterious
properties and third by disposing of the raw or treated
materials in such a way as to avoid public health and
aesthetic impact, where possible making use of nature to
further "treat" those organic components which can be degraded
by processes of decay. The problems of disposal have been
exacerbated by the nineteenth century practice of diverting
industrial aqueous wastes to the treatment facilities used
for human sewage, thus permitting effective breakdown of their
degradable components, but enhancing the contamination of
sewage by persistent or toxic materials.

DISPOSAL ROUTES OF SEWAGE SLUDGE
   2.   Sewage sludge production depends upon the size of the
contributing population, the degree to which sewerage is
provided and the level of treatment applied. The production
per unit of population may be seen as an index of the development of a country's sanitary system (Table 1).
   3.   The choice of a disposal route for sludge is related to
the availability of outlets and their environmental, social
and economic costs. Most of the sludge in Europe is disposed
of to land, either to sanitary landfill or for use in agricultural, forestry or park lands as fertiliser. The primary
constraint on this use of sludge is availability of suitable
land although there may also be constraints associated with
the sludge itself - its contamination by pathogens and
persistent toxic substances. Although controlled land
spreading is the most obviously beneficial use of sludge,

Table 1. Sewage sludge production in North Sea bordering countries and some other Oslo Convention signatory states

Country	Population (M)	% Pop'n sewered	% Pop'n served by treatment	Total Waste water load (mpe)1	% dis-charged to sewer	% sewage served by treatment	Current sludge production KT pa	Sludge production per head of population kg/yr
NORTH SEA BORDERING COUNTRIES								
UK	55.9	95	84	102	75	88	1500	27
FRG	61.7	86	66	182	62	70	78	36
Netherlands	13.9	40	79	31	68	86	230	17
Denmark	5.1	92	57	15	31	61	130	26
Belgium	9.9	55	22	31	52	24	70	7
Norway	4.1	55	25[2]	4.1[3]			55	13
OTHER OSCOM COUNTRIES								
France	53.6	56	45	146	38	74	840	16
Ireland	3.4	58	26	6	48	25	20	6
Luxembourg	0.4	76	55	1	83	80	11	26
Spain	37.0			373			45	1
Sweden	8.3	86		83			210	25

Notes: Source - recalculated from ref. 1
 1 including industrial waste water
 2 a further 41% is treated by septic tank
 3 domestic load only

this is often not looked on favourably because of the
association in the public mind of sewage and disease. Thus
alternatives are required, especially in densely populated
areas, and three possibilities are tipping, incineration or
disposal to sea. Tipping sewage sludge raises the question
of competition from the large and growing requirement for
land for disposing of other types of waste. Incineration has
substantial energy costs, can cause atmospheric pollution,
and has a high social profile, tending to rouse opposition
from residents in areas selected for plants.

4. Marine disposal also is constrained in most European
mainland countries by lack of ready access to coastlines and
by the sensitivity of local ecology. In addition, sea disposal is often politically unacceptable. Thus, of the
countries bordering the North Sea only Britain and the
Netherlands, both with a high coastal population density,
dispose of sludge to sea; elsewhere,Ireland and Spain, both
with long coastlines and extensive mountainous or wetland
areas also use the marine route. In the USA, coastal states
find sea disposal attractive because of the high population
density along the seaboard.

EUROPEAN TRENDS IN SEWAGE SLUDGE DUMPING AT SEA

5. Trends in marine dumping since the signing of the Oslo
Convention have varied between countries depending on basic
differences in philosophy on the use of the sea. The Federal
Republic of Germany dumped sludge in the German Bight from
Hamburg (ca 15-20,000 dry tonnes per year), but from 1980
this load was diverted to a point off the shelfbreak in the
south western approaches to the Celtic Sea, and in 1983 FRG
stopped dumping at sea. Heavy metal contamination of this
sludge prevented its use in agriculture and it is now
exported to'the German Democratic Republic for disposal.

6. In Ireland, about 200 kt wet weight of primary sludge
from the Dublin conurbation is dumped off Dublin Bay and this
figure has been rising in recent years as improvements are
made to the sewage systems. Digestion has been considered to
reduce inputs, but there are no satisfactory land based outlets for it.

7. In the UK, though most of the sludge is disposed of to
land (ref. 2), sea disposal has been looked on as an environmentally, socially and economically sensible disposal option
for sludges from large conurbations with access to the sea
and constraints upon land disposal. Sludges are dumped at sea
from sewage treatment works operated by most of the English
coastal Water Authorities and by the Strathclyde and Lothian
Regional Councils in Scotland. Thirteen disposal sites are
used around the British coast (Fig. 1). Disposal is dominated
by the input to the Thames (3 disposal sites), though it should
be noted that the tonnage is inflated by the unusually high
water content of this sludge (~98%). The Thames load peaked
at over 5 mt wet weight in 1979-81 and has thereafter declined

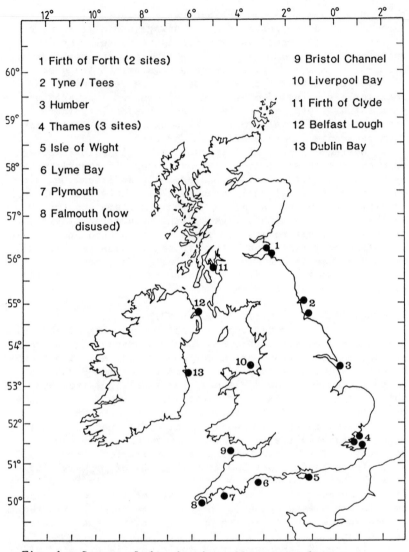

Fig. 1. Sewage sludge dumping sites round Britain
and Ireland.

slightly; it is likely to remain more or less stable in terms
of dry matter inputs (ca 100 kt) in the long term, though the
water content will decrease. Liverpool Bay receives over $1\frac{1}{2}$
million wet tonnes of sludge, and this may rise as the Liverpool sewerage schemes come on line. In the Garroch Head area
of the Clyde the disposal rate every year by the Strathclyde
Region is again $1\frac{1}{2}$ million wet tonnes. Disposal in the
Tyne/Tees area off the NE coast of England currently amounts
to around 0.5 Mt wet weight per year, but is rising as the
Tyne and Tees sewage treatment schemes develop, diverting

loads from the rivers to the open sea. Other sites round
Britain receive smaller quantities. The overall quantity
licensed for disposal (as opposed to dumped) from Britain
since 1979 has been and is likely to remain, roughly constant
at around 11.5 Mt wet weight.

NATURE OF THE MATERIAL

8. The material from the largest British operations,
namely the 4-5 Mt wet dumped annually at the Barrow Deep site
in the Thames and the 1.5 Mt dumped in Liverpool Bay, is
digested sludge. The remainder is undigested primary sludge,
derived from settlement of raw sewage. The gross differences
between primary and digested sludge are that digestion results
in an overall reduction in Biological Oxygen demand (BOD) and
total dry matter (ca 30%) as well as in the Chemical Oxygen
demand (COD) and the grease content. In addition, the
particle size is reduced and this may affect settling
characteristics (ref. 3). Digestion also leads to a diminu-
tion in pathogen levels and a loss of nitrogen from the
sludge (ref. 4) while the reduction in carbon content will
bring proportional changes in phosphorous and metal concentrations. The bulk of the metal contaminants remain bound
to very small particles (10 um) within the sludge (ref. 5)
and most organic contaminants will concentrate in the
particulate fraction.

9. Nutrient inputs from sewage sludge dumped in UK waters
of the North Sea in 1981 (Table 2) represent about 0.4% of
total nitrogen inputs and $2\frac{1}{2}$% of total phosphorous inputs to
the North Sea (ref. 7). Metal inputs from UK sludges (Table 3)
peaked in 1979/80 (the apparent rise before then partially
represents improvement in data gathering) and have since

Table 2. Nutrient inputs to the North Sea from sewage
sludge disposal in the UK (1981 data) (ref. 6)

		NITROGEN		PHOSPHOROUS	
		Concentration %	Max. input T/yr	%	Max. input T/yr
UK	Forth	3.7	353	0.5	189
	Tyne	-	798	-	225
	Tees	-	27	-	7
	Humber	2.1-5.1	338	0.7-2.2	93
	Thames (Barrow Deep)	7.5-7.8	9,210	1.5-1.8	2,125
	Harwich	(3.1-6.8)	727	(0.5-1.4)	150
	S. Falls		205		42
			10,142		2,317

Table 3. Inputs of metals in sewage sludge to English coastal waters (Tonnes)

	1976	77	78	79	80	81	82	83	84
Mercury	1.4	1.6	1.6	1.7	2.7	1.5	0.9	0.4	0.4
Cadmium	7.7	8.6	5.9	6.0	6.5	4.3	3.5	2.9	2.8
Lead	98	109	164	162	114	105	99	92	93
Copper	105	115	120	118	116	117	111	103	109
Zinc	457	485	608	604	392	367	254	278	188
Nickel	27	25	25	20	18	14	13	13	13
Chromium	56	69	74	64	61	53	50	47	47

levelled off or declined. The bulk of the English inputs are to the Thames and Liverpool Bay where considerable improvements have been achieved since the passing of the Dumping at Sea Act 1974; in particular, mercury inputs have been considerably reduced. Inputs to the Spurn Head dumping ground off the Humber have also sharply decreased in line with reduction in total inputs as well as improved sludge quality; this reduction has been offset by the recent rise in inputs to the dumping site off the Tyne. The sludges from the Tyne/Tees area are generally of good quality with respect to contaminants. In Scotland, a similar picture of a reduction in metal per unit of sludge is evident (Table 4). Much of the achievable reduction in contaminant levels in the UK has now occurred and though pressure by the licensing authorities on the sewage treatment agencies for further improvement continues, it is apparent that treatment is being applied with

Table 4. Weights of sewage sludge and associated metals dumped in the Firth of Clyde

Year	Sludge Dry Tonnes x 1000	Copper Tonnes	Lead Tonnes	Zinc Tonnes	PCBs Tonnes
1975	50	48	43	126	0.3
1981	50	48	52	76	0.04
1982	55	30	32	37	0.02
1983	80	35	37	49	0.02

diminishing returns. This is most noticeable in the case of copper and lead, where multiple diffuse sources contribute; further reductions in lead input are likely only as controls on lead in petrol and paint take effect. As a proportion of total inputs (Table 5) sewage sludge is a relatively small contributor of metals compared to river, atmospheric and

Table 5. Quantities of heavy metals entering the North Sea, 1981 data (T/yr), adapted from ref. 8

Metal	Hg	Cd	Pb	Cu	Zn	Ni	Cr
Sewage Sludge (all sources)	2	5	120	150	480	25	102
Total input	20–33	150–165	1780–3330	2430–2470	14000–14100	2130–280	1560–2000
Sewage sludge as % of total	5-9	3	4-6.5	5.5-6.5	3.5	1	5-6.5

dredge spoil inputs (ref. 8; ref. 7). It is worth noting that there have been improvements in UK sludge quality since 1981 (Tables 3 and 4), especially with respect to mercury, so that sewage sludge may now contribute proportionately lower inputs. Comparable data on inputs of man-made trace organics in sewage sludge are less readily available, but there is no reason to believe that the position would be proportionately different, and this is supported by the Clyde data (Table 4).

TOXICITY
    10. Sludges usually contain a wide range of materials other than the trace contaminants listed above, some of which, eg ammonia, may exert a toxic effect, especially considering the large quantities in which sludges are dumped. In reality, static tank toxicity tests (ref. 9) did not indicate any appreciable toxicity to a variety of adult fish and shellfish though larvae of shrimps (Crangon) were more sensitive. It was estimated (ref. 10) that in the normal post-dumping conditions in the New York Bight, dilution would bring the sludge well below the "no observed effect concentration" (NOEC) levels to the most sensitive Crustacea. In a 60 day bio-accumulation study to assess metal uptake (ref. 9) some intake of lead zinc and copper occurred with very highly contaminated sludges, but mercury and cadmium levels were lower than expected; it is possible that in sewage sludges, these metals are not readily available to marine organisms. However, field studies indicate that accumulation can occur in organisms in the vicinity of dump sites, underlining the importance of obtaining a good degree of dispersion to overcome both acute and chronic impact.

PATHOGENS AND PARASITES
    11. Sewage is a potent source of pathogens and parasites (ref. 11 and see Table 6), and this militates against the more extensive beneficial use of sludge in agriculture. Raw sewage discharged to sea has historically given rise to

Table 6. Pathogens present in sewage sludge
(adapted from ref. 4)

---

1. Bacteria, particularly <u>Salmonella</u> species
2. Viruses, particularly the Enteroviruses
3. Eggs of metazoan parasites, including tapeworms
   (<u>Taenia</u>) and roundworms (<u>Ascaris</u>, <u>Trichuris</u>).
4. Cysts of protozoan parasites (<u>Giardia</u>, <u>Acantha-
moeba</u>)

---

problems of contamination at coastal shellfish beds, leading
to closures of fisheries, and corresponding problems at
bathing beaches. Offshore disposal of sludge from vessels or
long sea outfalls at sites carefully selected to avoid shore-
ward transport of sludge materials, and out of contact with
significant fisheries helps to reduce this problem. Many
pathogens and parasites cannot survive in seawater, though
some (eg <u>Acanthamoeba</u> and many viruses) are sufficiently
resistant to be potentially useful indicators of the extent
of contamination of the sea bed by sludge (ref. 12).

CHOICE OF DISPOSAL SITE - MANAGEMENT OF NEAR-FIELD IMPACT
    12. The immediate potential problems at the point of dis-
posal are deoxygenation in the water column due to high BOD,
deoxygenation in the sediments by accumulation of excessive
organic matter, and acute toxicity from major components
(eg ammonia). A suitable disposal site will provide
adequate initial dilution to overcome these problems, and,
subsequently, sufficient dispersion to overcome chronic
effects due to local accumulation of parasites, pathogens or
toxic substances.
    13. At almost any waste discharge there is a local zone in
which the effects exerted by the waste are measurable.
Sensible management policy suggests first that the impacted
zone should be small with respect to the distribution of
significant local resources and should not overlap with shellfish beds or bathing beaches, and second that there should be
no significant consequences outside the zone. Normally this
leads to an approach of limiting impact in the local zone to
barely detectable biological change, though in the case of
piped discharges or dumping in accumulative sites, more
intensive changes in the biology, up to complete abiosis, may
be acceptable in a confined area if this limits the impact
outside the zone. An important final criterion is that,
having established the scale and extent of acceptable impact,
there should be no deterioration with time, which would
suggest inadequate assessment of assimilative capacity.
    14. Site selection is obviously constrained by geographic
location, and the rate and scale of disposal will have to be
suited to the local dispersal abilities and the degree of
local impact found acceptable. Most of the UK disposal sites
are strongly dispersive. In the Thames Barrow Deep site some

areas of accumulation do occur (ref. 13), with localised biological consequences, though tracer studies indicate that on the whole the large load is effectively dispersed (ref. 14). The Tyne and Tees sites are somewhat less dispersive and thus more limited in their capacity if significant local effect is to be avoided. The now disused site in the German Bight is also an area of weak tidal current and cohesive sediment (ref. 15) and the overlying waters are prone to stratification and subsequent bottom water de-oxygenation in summer. Studies (ref. 16 and 17) suggest an unstable local biology, in which the instabilities may have been exacerbated by sludge dumping as well as periodic summer low oxygen conditions and winter storms.

15. The site at Garroch Head in the Clyde Sea Area is a good example of an accumulating situation (ref. 18). The area is comparatively sheltered and the water relatively deep, about 80 metres and slow moving. The sludge sediments quite quickly to the bottom and accumulates in a restricted area so that a zone of about 15-20 km^2 can be identified as the area of major impact. The advantage of an accumulating site of this nature is that the effects are contained and monitoring operations can be accurately focused.

MONITORING OF SEWAGE SLUDGE DISPOSAL

16. The most significant effect of sludge is on the sea bed where sludge particles settle, and may include accumulation of carbon, pathogens and other microflora and persistent substances, deoxygenation of sediments and changes in benthic animal communities. The purposes of monitoring are to assess local impact at the disposal point, to measure temporal and spatial trends in this impact in both contamination and biological effect, and to determine the effect on public health of pathogens and persistent toxic contaminants in marine food chains.

17. The first two of these aims requires monitoring of conditions at and around the site ('near field'). The last is concerned with 'far field' effects at a distance in time and space from the dumping operation, where it will be difficult to distinguish sludge effects from those of other natural and anthropogenic inputs. This 'far field' monitoring of sludge disposal is an integral part of national and international efforts to monitor coastal water quality and the level of contamination of seafood products (ref. 19). The contribution of sewage sludge to such contamination cannot be assessed from field studies alone. Analysis of local input sources is also required, and research into the relative bioavailability of the substances in the different waste matrices. The organic matrix of sewage sludge, for example, may effectively bind several metals and their organic derivatives, reducing their immediate, and sometimes long term bio-availability (ref. 20).

18.   Near field studies may have two phases. Initially extensive spatial surveys are carried out before and in the first years after commencement of dumping to verify that the management aim of limitation of area of immediate impact has been achieved, and that the impact within this area is acceptably small. In this context it is useful to assess the distribution of benthic infaunal species and modern methods of data handling involving classification and ordination packages permit detection of very subtle changes in benthic community structure. However, sometimes the effect on the benthic community is too small to be readily measurable (ref. 21) and in these cases it is often possible to identify the impacted area on the bottom by altered sediments, enhanced organic carbon or trace metal levels, or by the presence of sludge derived tracers such as tomato pips and other vegetable remains (ref. 12). These approaches can also be used to identify the maximum spatial extent of deposition, which is normally greater than the area showing biological effect.

19.   Having established the spatial extent of impact, its variation from year to year should be examined. Such studies should be relatively intensive and are thus more limited in scale than the extensive spatial study. The determinants to be monitored may also be reduced and focused on contaminant levels in sediments or selected biota, or population change in a few components of the benthic community. It is advisable to check the spatial extent of the impact at intervals since this may be the only indication of changes, though normally an expansion of the margins would be associated with increases in intensity of effect at the centre.

20.   In Britain monitoring studies, both of spatial extent and temporal trend are carried out by the licensing authorities (MAFF and DAFS) and the licencees (Water Authorities, Regional Councils). Most of the UK disposal sites are dispersive and local effects are not very marked; an exception already referred to is the accumulative Garroch Head site in the Clyde where effects, though severe to the point of anoxia at the centre of the site, are tightly confined around the location of dumping (ref. 18).

THE FUTURE

21.   Sewage sludge will always be with us and constraints on its use and disposal on land are not decreasing, so the need for marine outlets is not likely to reduce despite political pressures against it (ref. 10; ref. 22). Three issues related to the future of sea disposal are changes in its acceptability, changes in disposal practices and development of beneficial uses at sea.

22.   Leaving aside a generalised opposition to dumping the two main problems are the toxic and the nutrient components of sludge. While considerable strides have been made in reducing the metals and xenobiotics, it is clear that further

effort is producing diminishing returns. It is essential that these components do not increase, and that their contribution to the sea from sludge is controlled. With regard to the nutrient content, even though advanced treatment will reduce BOD, the basic nutrients remain to be disposed of, and it is worth noting that the issue of eutrophication of coastal waters is a growing concern and the dynamics of nutrients in the sea is far from fully understood (ref. 23). In terms of public presentability, it is therefore again important that the contribution of sludge to coastal nutrient inputs is controlled to prevent unacceptable areas of enhanced algal production.

23. In Europe and eastern North America there has been a tendency to dump sludge relatively far offshore from vessels rather than to discharge through pipelines. The Hague outfall to the shallow coastal zone is expected to be terminated in the not too distant future, but long sea outfalls are in use on America's west coast which rely on very deep water (at the shelf edge) and good circulation close to the coast. The logical extension of this approach, especially as a "not in my backyard" concern for coastal areas grows, is to take sludge to the open ocean even if this is far offshore. Three options are apparent, dispersal into ocean surface waters, injection below the permanent thermocline or injection close to the sea bed. Surface disposal offers a high degree of dispersion and was used by the Federal Republic of Germany for up to 300 kt pa of wet sludge between 1981 and 1983 without measurable adverse effect. However, surface disposal, especially if not very far off the continental shelf, would need to be carefully sited to avoid onshore transport and contamination of shelf waters. Injection below the thermocline would reduce this possibility. Injection of the sludge to the ocean floor has considerable attractions. The main concern is with possible effects on the abyssal animal community, which might be altered by concentrated inputs of sewage materials to areas which normally receive much smaller levels of carbon input. Further research would be required before adopting this option.

24. In principle there would seem to be no reason why beneficial use could not be made of sludge at sea, as on agricultural land, to enhance productivity. In one sense this already occurs naturally, the productivity of coastal waters being increased by terrigenous run-off of nutrient and carbon-bearing waters. The importance of the Thames, including its sewage inputs, to productivity and fisheries in the Southern Bight has been recognised (refs 24 and 25) and the idea of enhancing natural productivity with deliberate fertilisation is not new. During the last war, extensive experiments were carried out (ref. 26) with the express purpose of artificial fertilisation in a Scottish sea loch. Although this work did not prove successful in its initial aim of raising flatfish yield, it did provide much useful

information on the effects of added nutrients, and the concept is again under discussion (ref. 27). However, in view of the lack of understanding of eutrophication processes, marine disposal of sludge could not yet be carried out with the precision of agricultural fertilisation, not least because it is difficult to calculate and control the dosage required. Nevertheless, the continued need for marine outlets for sludge suggests that research on coastal productivity and the role in it of nutrients, growth factors, trace metals and organic matter (all readily supplied by sludge) could yield long term beneficial results.

CONCLUSION
25. Marine disposal of sewage sludge by vessel or pipeline, properly managed, results in only minor local impact and, as currently practised, contributes only a small fraction of the total inputs of organic matter, nutrients and persistent contaminants to the North Sea. Sludge disposal is by no means the only route to the sea of sewage derived materials, which enter also via rivers and direct treated or untreated discharges. However these sources are more difficult to quantify, being often individually smaller in quantity, very much larger in number, and more diffuse than the large discrete sludge disposal operations discussed here. There is currently no satisfactory evidence of any other than local effects of sludge disposal operations apart from their relatively minor role in contributing to inputs of persistent substances.

REFERENCES
1. VINCENT A.J. and CRITCHLEY R.F. A review of sewage sludge treatment and disposal in Europe, 1982, Report 442-M, Water Research Centre, Medmenham.
2. HEALEY M.G. Guidelines for the utilisation of sewage sludge on land in the United Kingdom. Water Science and Technology, 1984, 16, 461-471.
3. GARBER W.F., OHARA G.T., COLBAUGH J.E. and RAKSIT S.K. Thermophilic digestion at the Hyperion Treatment Plant. J. Water Pollution Control Federation, 1975, 47, 950-961.
4. WOOD P.C. Sewage Sludge Disposal Options. In Kullenberg, G., (Ed). The Role of the Oceans as a Waste Disposal Option, 1986.
5. CHAPMAN D.V. The distribution of metals in sewage sludge and their fate after dumping at sea. The Science of the Total Environment, 1986, 48, 1-11.
6. GROGAN W. Input of contaminants to the North Sea from the United Kingdom (Report for the Department of the Environment). Institute of Offshore Engineering, Heriot-Watt University, Edinburgh, 1984. Mimeo, 203 pp.
7. NORTON R.L. The assessment of pollution loads to the North Sea. Technical Report No. TR 182, 1982, Water Research Centre, Medmenham.

8. HILL J.M., MANCE G. and O'DONNEL A.R. The quantities of some heavy metals entering the North Sea. Technical Report No. TR 205, 1984, Water Research Centre, Medmenham.

9. FRANKLIN F.L. Laboratory tests as a basis for the control of sewage sludge dumping at sea. Marine Pollution Bulletin, 1983, 14, 217-223.

10. FAVA J.A., McCULLOCH W.L., GIFT J.J., REISINGER M.J., STORMS S.E., MACIOROWSKI A.F., EDINGER J.E. and BUCHAK E. A multidisciplinary approach to assessment of ocean sewage sludge disposal. Environmental Toxicology and Chemistry, 1985, 4, 831-840

11. DAVIS R.D. Control of contamination problems in the treatment and disposal of sewage sludge. Technical Report TR 156, 1980, 79 pp, Water Research Centre, Medmenham.

12. VIVIAN C.M.G. Tracers of sewage sludge in the marine environment: A Review. The Science of the Total Environment, 1986, 53, 5-40.

13. NORTON M.G., EAGLE R.A., NUNNY R.S., ROLFE M.S., HARDIMAN P.A. and HAMPSON B.L. The field assessment of effects of dumping wastes at sea: 8. Sewage sludge dumping in the outer Thames Estuary. Fisheries Research Technical Report, 1981, No. 62. Directorate of Fisheries Research, Lowestoft.

14. TALBOT J.W., HARVEY B.R., EAGLE R.A. and ROLFE M.S. The field assessment of dumping of wastes at sea: 9. Dispersal and effects on benthos of sewage sludge dumped in the Thames Estuary. Fisheries Research Technical Report, 1982, No. 63, Directorate of Fisheries Research, Lowestoft.

15. NORTON M.G. and CHAMP M.A. The influence of site-specific characteristics on the effects of sewage sludge dumping. In Duedal, I. and others (Eds). Wastes in the Oceans Series John Wiley (In Press).

16. CASPERS H. Longterm changes in benthic fauna resulting from sewage sludge dumping into the North Sea. Water Technology, 1980, 12, 461-479.

17. RACHOR E. Faunenverarmung in einem schlickgebiet in der Nahe Helgolands. Helgolander Wissenschaftliche Meeresunters, 1977, 30, 633-651.

18. MacKAY D.W. Sludge dumping in the Firth of Clyde - a contaminant site. Marine Pollution Bulletin, 1986, Vol. 17, pp 91-95.

19. ANON. The ICES Co-ordinated Monitoring Programme for Contaminants in Fish and Shellfish, 1978 and 1979, and six-year Review of ICES co-ordinated Monitoring Programme. Co-operative Research Report, 1984, No. 126, International Council for the Exploration of the Sea, Copenhagen, Denmark.

20. BRYAN G.W. Bioavailability and Effects of Heavy Metals in Marine Deposits. In: Ketchum, B.H. et al. Wastes in the Ocean. Vol. 6, 1985.

21. EAGLE R.A., HARDIMAN P.A., NORTON M.G. and NUNNY R.S. The Field Assessment of Effects of Dumping Wastes at Sea: 4. A survey of the Sewage Sludge Disposal Area off Plymouth. Fisheries Research Technical Report, 1979, No. 50,

Directorate of Fisheries Research, Lowestoft.
22. WALKER D.L. Sludge Disposal Strategy: A UK viewpoint.
In Dart, M.C. and Jenkins, S.H. (Eds). Disposal of Sludge
to Sea. Water Science and Technology, 1981, Vol. 14 (3),
127-136.
23. ANON. Report of the ICES Special Meeting on the Causes,
Dynamics and Effects of Exceptional Marine Blooms and
Related Events. ICES Doc. C.M. 1984/E:42, International
Council for the Exploration of the Sea, Copenhagen, Denmark.
24. GRAHAM M. Phytoplankton and Herring, Part II, Distribution of Phosphate in 1934-36. Fisheries Investigations,
London, Series II, 1938, 16, No. 3.
25. CARRUTHERS J.N. Seafish and residual sewage effluents.
Intelligence Digest Supplement (October issue), 1954.
26. GROSS F., NUTMAN S.R., GAULD D.T. and RAYMONT J.E.G.
A fish cultivation experiment in an arm of a sea-loch, I-V,
1950, Proc. Roy. Soc. Edinb. B64, 1-135.
27. SEGAR D.A. et al. Beneficial use of municipal sludge
in the ocean. Marine Pollution Bulletin, 1985, vol.16, 186-
191.

9. Marine disposal of sewage sludge by North West Water Authority and Strathclyde Regional Council

E. HARPER, BScTech, MIWPC, MIPHE, Chief Scientific Adviser, North West Water Authority, and W. T. GREER, BSc, FICE, FIWPC, Director of Sewerage, Strathclyde Regional Council

SYNOPSIS. North West Water Authority and Strathclyde Regional Council are dependent on marine disposal of sewage sludge by vessel, presently for 1.9 and 1.69 million wet tonnes respectively per annum. The sites used are contrasting ones, in Liverpool Bay strong currents ensure rapid dispersion, whereas Garroch Head is an accumulating site. An outline of the historical background, the environment studies, and some reasons why marine disposal is considered the best environmental option are given by each author for his authority.

PART I - NORTH WEST WATER AUTHORITY PRACTICE

HISTORICAL
1. Sewage sludge has been disposed of into the Irish Sea by vessel from the cities of Manchester and Salford since about 1895. The disposal area has been subject to constraints from the earliest days when regulations were imposed "under which sludge can be deposited at sea under the same conditions as those which have been enforced for many years with respect to the disposal of the Liverpool Town refuse." by the then Acting Conservator of the Mersey in 1896 and remained valid until the Dumping at Sea Act 1974: these regulations were:-
i) The refuse shall be tipped outside the North West Light Ship at a depth of not less than 15 fathoms.
ii) The refuse shall be deposited as near as possible after high water.
iii) The masters of the vessels carrying the refuse shall signal by whistle the times of passing the North West lightship, and note the time in the log.
iv) Extracts from the logs of vessels shall be furnished monthly to the acting conservator.
2. Liverpool Corporation ceased dumping town refuse in 1953 and the North West Light Ship no longer exists. The disposal area, now defined more precisely by the Ministry of Agriculture, Fisheries and Food, (MAFF) remains in the same vicinity some 30 km west of Formby Point near Liverpool.

The site is positioned in a highly energetic zone and has proved to be highly dispersive. It is as recently as 1972 that it was possible to write a report on this disposal route under the title "Out of Sight Out of Mind" (ref 1) confident that the title was apt. The authors said that it reflected "our view of the disposal to sea of sewage sludge by the Manchester and Salford Corporations". The last few years have seen the growth of protectionist attitudes to the marine environment which seek to protect the seas ahead of land, inland waters and the air. A report on the practice of sea disposal would more aptly now be entitled "In mind, with the Greens in view".

A MIXED SLUDGE STRATEGY

3. Before discussing North West Water Authority's (NWWA's) use of the sea for sludge disposal it is necessary to consider this route in the context of the established practice of the Authority of a "mixed sludge strategy". This term is used to describe the adoption of various routes to dispose of the total sludge production, with the preferred route being selected after balancing a number of factors. These factors include the type of sludge; the degree of sludge treatment; the possible environmental effects on land, rivers, sea and air; physical constraints such as vehicle movement and storage sites; security of service and of course the costs of alternative routes. The route selected is decided with the prime aims of disposal of sludge with least risk of adverse environmental effect, at least cost. These two aspects are equally crucial and consideration of one without the other is considered unacceptable within NWWA practice. The prime purpose of the sewerage system is to remove the wastes carried by the water carriage system away from homes in order to protect public health, and that of sewage treatment to protect the aquatic environment from the consequences of this action. Sludge is an inevitable by-product of sewage treatment and to use a route which runs counter to the prime purpose of sewage treatment would clearly be self defeating.

4. The NWWA relied on the following routes during 1984/85 to dispose of the sludge arising from sewage treatment works serving some 7 million people:

Table 1. Quantities of sewage sludge for disposal

Disposal route	Quantity disposed 1984/5 (tonnes dry solids)	
Sea	54,000	(44%)
Agricultural land	36,000	(29%)
Other land	15,000	(12%)
Increase in store	13,000	(11%)
Incineration	5,000	(4%)
TOTAL	123,000 tonnes dry solids	
	equivalent to 1,900,000 wet tonnes	

5.   This quantity represents only about 80% of the total raw sludge produced as about 30,000 tonnes are destroyed by treatment using anerobic heated digestion.   An increase of around 20,000 tonnes dry solids of digested sludge per year is expected to arise by the early 90's from first time sewage treatment for discharges to the Mersey Estuary from a contributing population of over 1 million.

6.   Quantities to Sea Figure 1 shows the quantities discharged to sea from the beginning of this century expressed as dry solids (tonnes per year). (ref 2 updated).
7.   The large quantities in earlier years are the result of the chemical precipitation methods employed at Salford up to 1950, and also similar treatment at Manchester in the early part of the century.   The decreased annual quantities for the war years and immediately afterwards were due to the sinking of the first SS "Mancunium" with extensive use of emergency land disposal.   The gradual increase in subsequent quantities reflects the greater numbers and improved performance of sewage works.   The fleet has in recent years comprised of four purpose built vessels, two approximately 1500 wet tonnes carrying capacity, and two 3,000 tonnes both built in the early 70's.   One smaller vessel has recently been sold and there will be a further review when construction of pipelines will reduce the dependence on the Manchester Ship Canal.
8.   By the mid 60's it was clear that the increasing amount of sludge expected from new sewage works planned or being built as a result of improved pollution control legislation would exceed the on-site disposal facilities. These needs led to a suggestion of joint action by local authorities to consider the feasibility of pumping the sludges to either Manchester or Salford for sea disposal. This suggestion was followed up and aroused considerable interest amongst some 50 and possibly up to 100 local authorities in South Lancashire and North Cheshire with similar sludge problems.
9.   It was against this background that a working party on sludge disposal in Liverpool Bay was formed by the then Ministry of Housing and Local Government and began the work which lead to the series of reports 'Out of Sight Out of Mind' (ref 1).   This historical background is quoted so that the pressures on sludge disposal routes in the North West can be better appreciated and also it was the spur for much scientific work to be undertaken.   It was then expected that the quantity of sewage sludge disposed of to Liverpool Bay would increase six-fold over the period 1970-1975, this did not happen.

ENVIRONMENTAL STUDIES
10.   A broad description of the various efforts will be given and it can be estimated that this work, carried out since the late 1960s would cost well over £1 million in 1987

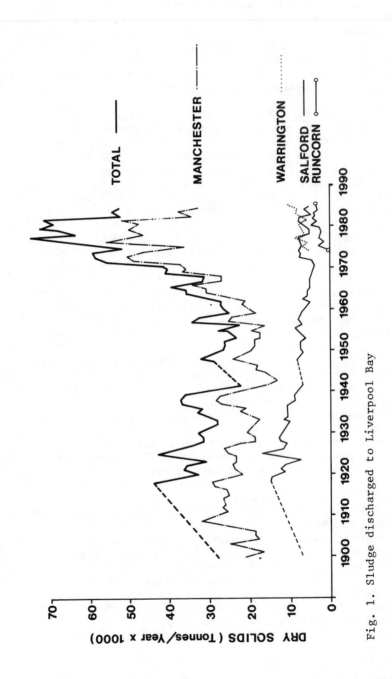

Fig. 1. Sludge discharged to Liverpool Bay

pounds. It is impossible to do justice and summarise all the work in a few lines but the author believes that few involved over the years would dispute the following general conclusions:-

(a) It is still very difficult to separate the effects of sewage sludge disposal from the many other sources of contaminants in the area, especially the River Mersey.

(b) There are now data using the long term records and statistical analysis which indicate a probable change in the content of certain metals in the silt fraction of sediments due to sewage sludge disposal, but not to demonstrate any apparent adverse effects.

(c) Changes in the benthic community have been detected over the years but direct correlation with sewage sludge disposal, if it exists, remains elusive.

(d) The quality of fish and shellfish has been affected by inputs of metals and other substances and sewage sludge is a relatively minor input of such materials but it may make a significant contribution through some route as yet unknown.

QUALITY OF SLUDGES FOR MARINE TREATMENT

11. There are four aspects of particular interest, the organic content, the microbiological content, nutrient enrichment and the content of persistent materials such as metals.

12. There can be little argument when the alternatives are considered that for the disposal of organic matter and reducing the microbiological content before there is potential contact with man, marine treatment is much more effective than land disposal or inland treatment. This conclusion is obvious when it is understood that even full biological sewage treatment will reduce the bacterial content for example of the indicator organism E.Coli, by 90-95%. This sounds highly effective but E.Coli counts will remain at perhaps 100,000 to 1 million per 100 ml and these have to be disposed of into rivers with perhaps 10 times or less dilution. With sludge disposal to sea the 10% removed is probably approaching the maximum which may be accumulated in the sludge and is discharged into a hostile environment with large dilution factors and where bacterial die-off is fairly rapid. In NWWA all sludges will soon be digested before sea disposal and this process will reduce coliform bacteria numbers by perhaps 90% as compared to those in raw sludge.

13. There have been studies over several years of the possible contribution made by sewage sludge to nutrient enrichment and in particular Phaeocystis growth which can reach nuisance proportions in the Irish Sea. No link has been established and nitrogen input is certainly relatively small from this source, although phosphorus may be significant.

14. The potential problem with disposal to the marine environment is not the organic matter which is degraded naturally or even the microbiological content, but the legitimate concern that non-degradable materials and persistent chemicals may accumulate and pass through the food chain, damaging the ecosystem and eventually reach man. In NWWA there has been particular concern and action over the levels of mercury in the sludges and in direct industrial discharges to the marine environment. Figure 2 shows the results of a successful campaign, with the full cooperation of industry, to reduce the discharges of mercury to the environment. Superimposed it shows the mercury levels in fish muscle tissue mainly a "shopping basket" of mixed species of fish which MAFF use as indicative of dietary intakes (ref 3 updated). It is this latter figure which is taken as a measure of achievement of the EEC standard of 0.3 mg per kilogram wet weight.

15. The mercury story is a complex one and the exact route from mercury inputs to fish flesh is not known. It seems likely that as there has been fluctuations in fish flesh levels between 0.25 and 0.29 mg per kilogram for over 10 years - despite a 90% reduction in industrial discharges and a similar reduction in mercury in sewage sludge discharges - that the path way is more associated with the mercury tied up in the silts of the Mersey Estuary and Liverpool Bay rather than present inputs. This accumulation stems from decades of mercury discharges, with perhaps a contribution from dredging disposal which redistributes the contaminated silt. The substantial improvement in the quality of fish flesh in the early 70's, confirmed from different sets of data may well have been associated with reductions in organic mercury from both paper-mill discharges and from an industrial process utilising organic mercury as a catalyst. These contributions are impossible to quantify accurately retrospectively but the coincidence of a reduction in inputs at the relevant time could well be significant. The fact of the matter is that the fish flesh, although elevated, does meet the environmental quality standard; there can be few other areas of the law where there is so much concern expressed given that the requirements of the law are consistently met with about a 10% margin.

CONSEQUENCES OF CESSATION OF MARINE DISPOSAL ON NWWA

16. The consequences of ceasing this route have been examined. In the short term there would have to be increased road tankering of liquid sludge for application to agricultural land; intensified use of drying beds and recommissioning of disused drying beds with dried cake being conveyed by lorry for land reclamation or to tip; intensified use (24 hours per day) of operational filter plate presses and recommissioning of disused presses, with cake being conveyed by lorry to agricultural land or tip;

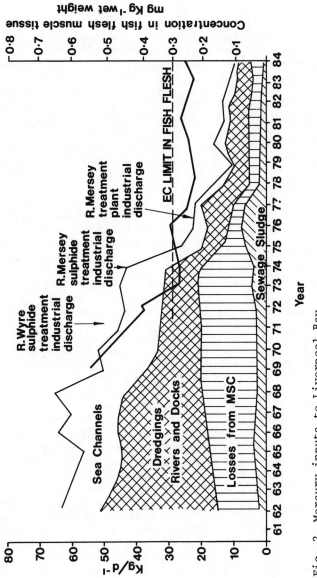

Fig. 2. Mercury inputs to Liverpool Bay

and use of spare capacity in the existing incineration plant. In broad terms about half of the sludge which is going to sea could be handled by these methods. The devasting problem which would remain would be how to dispose of the remaining 900,000 wet tonnes per annum.

17. The costs of these expedients would approximately double the existing sludge costs but of greater consequence the environmental impact would be enormous. The additional use of drying beds would undoubtedly cause widespread complaints of smell; this is one of the contributory reasons for phasing out many of them over the years and NWWA's preference for sea disposal for sludges arising from treatment works in built-up areas. None of the sewage works which presently send sludge to sea has filter presses therefore much of the liquid sludge would have to be conveyed by tanker to the works with filter presses. This would involve about 140 vehicle movements each working day. There is little tipping space available near the conurbations for sludge cake and this could lead to stockpiling, with adverse environmental impact. There could be up to 500 vehicle movements for tankering liquid sludge from the 27 sewage works which send sludge to sea. There is little agricultural land in the conurbations to receive the 1,900,000 wet tonnes per annum. The transportation by vehicle has been a sensitive topic in built-up areas in the North West from many years causing numerous complaints of noise and danger to pedestrians and to children. Much sludge would have to be stored in lagoons and this is a wasteful, objectionable and short-term measure which sterilises land and causes potential dangers to the public. The Authority has spent many years seeking to phase out such means of sludge handling. The quantities of sludge produced in the North West are immense. The 1,900,000 wet tonnes of sludge would cover a hundred football pitches to a depth of over 2 metres, - this is produced every year.

18. The Authority did in the late 70s seek to build an incinerator in the Croal Valley near Bolton with a capacity for dealing with 628,000 wet tonnes, largely to reduce the dependence on the sludge to sea which at that time was routed entirely along the Manchester Ship Canal. In the decision letter refusing the appeal by NWWA for planning permission the different methods of disposal were discussed and the comment that "sea disposal continued to be a satisfactory method" was made. Since that date and influenced by the decision the Authority has invested over £23 M capital in pipelines to reduce the dependence on the Manchester Ship Canal, to reduce tanker movements, and reduce long-term costs. It has also been estimated that to secure a site, to obtain planning permission and to design and construct a major incinerator would have a lead time, if indeed it was possible, of at least 7 years.

19. The projected increases in sludge to be disposed of
to sea are from treatment of crude sewage now discharged to
the Mersey Estuary. Can it be wrong to prevent the fouling
of shorelines near major centres of population and have the
equivalent of a pipeline 50 kms long discharging in a highly
dispersive area?

20. Can there be any doubt that marine disposal of sludges
in appropriate cases is the best environmental option?

PART II - STRATHCLYDE

HISTORICAL

21. The shipping of sewage sludge to sea in Strathclyde
was commenced by the former Corporation of Glasgow in 1904
from its second sewage works. This works was situated on
the north bank of the River Clyde at Dalmuir, 13 kilometres
downstream of the City Centre and then several kilometres
beyond the City's western boundaries. Glasgow's original
sewage works, which served the north eastern section of the
City had been opened ten years earlier at Dalmarnock.

22. Since Dalmarnock had been situated on a stretch of the
River Clyde which could only be reached by very small
vessels after they had passed through a tidal weir, its
sewage sludge was initially dried and pressed, thence sold
as "Globe" fertiliser. It was found from the earliest days
that this system of sludge disposal created problems. These
related to cost, difficulty of producing a satisfactory
product, labour troubles and the inability to dispose of it
continuously throughout the year.

23. The City Fathers, led by the Lord Provost, inspected
the London sea disposal operation in 1898 and reported back
favourably to the Corporation. Sea disposal of sewage
sludge was accordingly adopted for the remaining two Glasgow
sewage works at Dalmuir and Shieldhall, both of which are
located several kilometres downstream of Dalmarnock on the
navigable stretch of the River Clyde.

24. By 1914 it was decided that more secure sludge
disposal arrangements were required for the original sewage
works at Dalmarnock. A 9 kilometre long, 225 mm diameter
cast iron sludge pipeline was laid across Glasgow to
Shieldhall Sewage Works. Initially only surplus sludge
which could not be pressed was pumped from Dalmarnock.
Gradually the quantities increased however and since 1935
the total sludge produced there has been pumped to
Shieldhall for sea disposal.

25. During and immediately after the First World War, the
lack of artificial fertilisers led to a considerable demand
for dried sewage sludge. A second sludge pressing facility
was accordingly constructed at Shieldhall and the
Corporation's original sludge vessel was sold in December
1920. The second pressing plant hardly produced any sludge

cake however as the demand for the product rapidly fell away. Another ship had to be hurriedly constructed to replace the one which had been sold.

SLUDGE DISPOSAL GROUND

26. The original area for disposing of spoil in the Firth of Clyde was in deep water off the mouth of Loch Long, some 37 kilometres from Glasgow. An action in the Court of Session in Edinburgh in 1892 had however forced the Clyde Navigation Trust to take the dredgings from the Port of Glasgow, which were then fouled with sewage solids, to the present disposal ground south of the Isle of Bute (ref 4). The area used is beyond a three mile limit on the seaward side of Garroch Head which is the southernmost tip of the Isle of Bute.

27. In that area, the Firth of Clyde widens out to give a stretch of partially sheltered water approximately 17 kilometres wide between the Isle of Arran and the mainland coast of Ayrshire. The depth to the sea bed in this area is between 70 and 100 metres and the currents above it are fairly weak being less than 0.5 knots. The Strathclyde sludge disposal area is accordingly regarded as an accumulating disposal area.

28. In earlier years, the sludge vessels attempted to distribute their cargoes fairly widespread south of the three mile limit line. This changed however, as soon as the Regional Council obtained its first licence under the 1974 Dumping at Sea Act, since the licensing authority, the Department of Agriculture and Fisheries for Scotland, had decided that it would be preferable to concentrate the sludge within a circule of one nautical mile diameter. "Precision" sludge disposal has accordingly taken place since then.

29. It is of interest to record that in 1941, following the sinking by enemy action of Manchester's sole sludge vessel, one of the Glasgow ships, the original S.S. Shieldhall, was sent to Davyhulme and stayed there until 1947. During these six years, the remaining sludge vessel disposed of her cargoes in the original disposal area off Loch Long to enable the Glasgow service to be maintained. This disposal area is some 32 kilometres nearer Glasgow and there is no record of any adverse reports of unsatisfactory conditions having been experienced then.

GROWTH OF SEA DISPOSAL SERVICE

30. When only Dalmuir Sewage Works was dispatching sewage sludge to sea a the beginning of this century, the quantity disposed of annual amounted to around 200,000 tonnes. This increased to about 550,000 tonnes after Shieldhall Sewage Works was opened in 1910 when a second sludge vessel had been purchased and brought into operation.

31. By the outbreak of the Second World War, when all the sludge from Dalmarnock was also being sent for sea disposal and after the City of Glasgow had expanded its population and improved the sanitary arrangements of its citizens, the annual tonnage of sludge sent to sea had increased to over three quarters of a million tonnes.

32. Neighbouring authorities which had by then developed their own sewage works, looked jealously at the Glasgow Sea Disposal Operation. By August 1939, an agreement had been entered into with the neighbouring Burgh of Paisley whereby a sludge pumping main was to be constructed from its new sewage works to Shieldhall. It was not until 1971 however, that this scheme was constructed and brought into operation.

33. In 1962 the smaller neighbouring burgh of Barrhead sought assistance from Glasgow after its sludge heat treatment plant had irrevocably broken down. The Corporation agreed to allow Barrhead's sewage sludge to be brought to Shieldhall by road tanker. Numerous requests followed from other adjacent Burghs and County Councils. The Corporation agreed that the development of such a service was in the public interest and made suitable facilities available to these authorities also. Before the formation of Strathclyde Regional Council in 1975 accordingly, a regional Sea Disposal Service for sewage sludge was already in operation.

34. Now that Strathclyde is the sewerage authority for Glasgow and a 13,800 square kilometres area around it, the Region's policy has been to adopt sea disposal of sewage sludge for all sewage works within economical distance of Shieldhall for road transport. The fringe areas in Ayrshire, South Lanarkshire and Argyll and Bute still dispose of sewage sludge to agricultural land, tips, or via outfall sewers to the sea. Altogether, the sewage sludge which is now dispatched to sea from Shieldhall and Dalmuir Sewage Works in addition to that originating in these works, is brought in from over 40 other Strathclyde sewage works by a fleet of road tankers conveying a total of about 500 loads per week and also via two sludge pumping mains. The three Glasgow Sewage Works produce 1.16 million tonnes of sludge annually, to which is added about 530,000 tonnes from other sewage works throughout central Strathclyde, giving the total quantity disposed of at sea annually of 1.69 million tonnes.

35. The service is now being operated by the fifth and sixth purpose built sludge vessels. The present vessels, m.v. Garroch Head and m.v. Dalmarnock, can carry 3,500 and 3,100 tonnes of liquid sludge per trip respectively. This contrasts with the 1,000 and 1,5000 tonnes cargoes which the first two ships could carry.

SLUDGE QUALITY

36. Samples of sewage sludge are taken regularly and analysed by the Regional Council's Chemist and Public Analyst's Department. The results obtained are typical of those for sewages predominantly of domestic origin and compare well with the other United Kingdon sewage sludges which are disposed of at sea.

37. Only a very small percentage of the sludge is digested before dispatch to sea. Some digested sludge from the smaller works at some distance from Glasgow is shipped on farmland. The Region's five largest sewage works, viz, Dalmarnock, Shieldhall, Dalmuir, Paisley and Daldowie, dispose of all their sludge at sea. Full biological treatment is now carried out at all of these works, with the exception of Dalmuir. Accordingly, substantial quantities of surplus activated sludge are sent to sea as well as sludges settled out in primary settlement tanks.

38. In the 1960's officials of the Clyde River Purification Board advocated the sea disposal of difficult liquid wastes and sludges of industrial and manufacturing origin. The Corporation of Glasgow however, refused to deviate from its policy of only dealing with sewage works sludges despite the fact that another English Authority had provided a special corrosion resistant tank in one of its latest sludge vessels. I understand that this latter operation did not last long before the facility was withdrawn.

39. The quality of the Strathclyde sewage sludge has improved in recent years due to a considerable extent to the closing down of numerous factories. Although this has been welcomed by environmentalists, politicians at least have regarded this development as a very mixed blessing.

MONITORING OF SLUDGE DISPOSAL AREA

40. Until the 1974 Dumping at Sea Act licences were introduced, the Garroch Head sludge disposal area was protected by the normal quality controls applied to sewerage systems, particularly with respect to trade effluent discharges. Little investigatory work was then carried out at the actual sludge disposal areas. Studies were nevertheless carried out by both the Clyde River Purification Board and the Department of Agriculture and Fisheries for Scotland in the late 1960's and these revealed little obvious harm to the environment (ref 5).

41. Following the implementation of the Dumping at Sea Act in 1974, Strathclyde Regional Council engaged the Scottish Marine Biological Association, based at Dunstaffnage, near Oban, to conduct annual monitoring examinations of the sludge disposal area south of Garroch Head. Altogether, eight of these annual examinations have now been carried out.

42. The monitoring has consisted of sampling at about 60 stations, forming a grid covering an area of approximately 100 square kilometres, centred on the licensed dumping area. Core and grab samples of the sea bed are obtained at these stations. The samples are analysed to determine their Redox potential, organic carbon, heavy metals and organochlorine contents. Macrobenthic population determinations are also made by examining grab samples and collecting, enumerating, weighing and identifying the organisms present in these samples.

43. The disposal area is also trawled and specimens of demersal fish are examined for the presence of pathogenic bacteria and any abnormalities. The depth of the sludge deposit in the centre and at the edge of the disposal area is also determined from long core samples by physical examination and analysis of carbon levels down the core.

44. Negative redox values indicate reducing conditions immediately around the disposal area extending to about 10 square kilometres. There is an area of low redox values (below -100 mV) in the centre of the disposal area covering about 3.6 square kilometres. The bottom water immediately over this part of the sea bed is however fully oxygenated.

45. The sediments in the centre of the disposal ground have high carbon values. The metal concentrations do not appear however to be increasing and there is no evidence of any consistent differences between metal concentrations found in organisms taken at the disposal ground and those collected in a clean reference area some eight kilometres away.

46. There is also no evidence of contamination by human pathogens in the specimens of live fish taken from the area. High biomass and numbers of a few species of worms are found in the centre of the disposal area. The diversity of species increases with distance from the centre until communities characteristic of unaffected sediments are found about 3 kilometres away.

47. Core samples taken from the centre of the disposal area suggest that the maximum depth of the sludge layer is now about 20 centimetres. A fuller account of the annual monitoring surveys and their significant can be found in the evidence presented by Dr.Pearson of the Scottish Marine Biological Association to the House of Lords Select Committee on the European Communities when it was considering the Dumping of Waste at Sea (ref 6).

48. The benefits of an accumulating site such as Garroch Head, as opposed to dispersal sites are now becoming recognised more widely in contrast to the previous disperse and dilute approach which was formerly preferred. There is certainly a strong argument that it is more desirable to contain deposits in a limited sacrificial area than to scatter the material thinly over a much wider area and not really know its eventual fate (ref 7).

ACCEPTABILITY OF PRESENT SLUDGE DISPOSAL ARRANGEMENTS

49. Although the past decade or so has seen a great increase in the interest taken by scientists in the marine disposal of sewage sludge, there has not been a corresponding increase in the interest taken by the general public. In over 30 years of close association with the Garroch Head disposal operation, I have yet to receive a complaint from the public about this practice and there is no record of previous public concern. This record of over 80 years of trouble-free operation ought to be taken into account whenever the future of this service is under consideration.

50. It should also be borne in mind that due to war emergency conditions, the sewage sludge from Glasgow was actually deposited in a more land-locked area some 32 kilometres nearer the City. Again there is no record of public concern nor of resulting unacceptable conditions.

51. The Strathclyde operation has been accompanied by the development of a very active social use of the sludge vessels. Beginning on an ad-hoc basis during the first World War when parties of wounded soldiers were taken down river, the conveyance of parties of up to 70 passengers per trip has become an established tradition in the West of Scotland. M.V. Garroch Head has been provided with a high standard of accommodation for these passengers, who are now predominantly senior citizens.

52. Altogether, over 3,000 passengers are conveyed annually and this has generated immense good will towards the sea disposal operation. Indeed, during two periods in the late 1960s and mid-1970s, when awaiting delivery of new larger sludge vessels, these trips had to be temporarily discontinued. The public and political pressure to have them restored as soon as possible was very considerable and sustained.

53. The sea disposal service has undoubtedly stood the test of time. It provides a very positive method of disposal, even during adverse war time conditions, which is not dependent on gaining continuous access to farmland and remaining acceptable to farmers. Over the years in Strathclyde sludge pressing and heat treatment plants have been found to be unsatisfactory in practice. In addition, it is becoming increasingly difficult to obtain sites for sewage treatment facilities including even sewage pumphouses. The construction of large land based sludge disposal facilities such as incinerators or compost plants would almost certainly lead to widespread public protest.

54. Finally, the attractive cost of sea disposal of sewage sludge cannot be ignored by responsible public authorities. At present it costs just under £1 per wet tonne for shipping Strathclyde's sewage sludge to sea and this remains exceedingly competitive when compared to other possible alternatives.

REFERENCES

1. Out of Sight Out of Mind. Report of a working party on disposal of sludge to Liverpool Bay. Vol's 1-4. HMSO London 1972, 1973, 1976.
2. P.C.HEAD. Sewage Sludge Disposal in Liverpool Bay. Research into Effects 1975 to 1977 Part 2. DOE London 1984.
3. M.G.NORTON et al. Fisheries Research Technical Report Number 76 MAFF, Lowestoft 1984.
4. RIDDELL J.F. Clyde Navigation, J.Donald, Edinburgh 1979.
5. D.W.MACKAY and G.TOPPING Effluent and Water Treatment Journal, November 1980. pp 641-9.
6. HOUSE OF LORDS 17th Report of Select Committee on the European Communities - Dumping of Waste at Sea, HMSO 1986.
7. T.H.PEARSON Disposal of Sewage Sludge in Dispersive and Non-Dispersive Areas, Contrasting Histories in British Coastal Waters - NATO Workshop, Portugal.

Discussion on Papers 8-9

MR I. D. BROWN, Grampian Regional Council
The Authors of Paper 8 addressed an aspect of sludge
disposal which has had rather more than its fair share of
criticism from the environmental lobby. From my own
experience in Grampian Region I know only too well the
resistance by most farmers to accept sludge on land, rightly
or wrongly (and it is nearly all domestic sludge). A lot of
it in Grampian Region goes out to sea via long sea outfalls,
short sea outfalls and even shorter sea outfalls. This is
not a particularly satisfactory solution, but then I cannot
afford a sludge vessel, and even if I could, it would not be
much use in view of the number of small and widespread
communities.

On the assumption that shipping sludge to sea is the 'best
practical environmental option', one has to examine the
reasons for opposition to this practice. Two main problems
are identified in Paper 8: the toxic and the nutrient
factions of sludge. It would appear that reducing the toxic
component even further is increasingly difficult and
expensive - also, the eutrofication of coastal waters is of
growing concern. What then is the next step? Is it to ship
sludge to the open ocean as is suggested? Would the Authors
say what the likelihood of this happening is in the
foreseeable future and do they think this would satisfy
environmental objections?

Both North West Water Authority (NWWA) and Strathclyde
Regional Council or their predecessors have been shipping
sewage to sea since the turn of the century and one would
have thought that substantial pollution problems would have
become manifest by this time. However, public awareness of
our environment has increased since then, particularly in
the past decade. This is to be welcomed if, at the same
time, the desire for improvement is matched with the
resources to fund higher standards, which brings me to a
question to Mr Harper. NWWA's new Mersey Valley sludge
pipeline must now be operational - or nearly so - and this
will bring sludge to the Liverpool terminal for disposal to
sea from inland towns such as Bury, Bolton, Oldham and
Rochdale and, while one must congratulate the authority on

this engineering feat, can I ask what effect this is likely
to have on the total amount of sludge to be dumped and its
constituents? Also, what steps have been taken to keep out
list 1 substances, principally mercury and its compounds,
and reduce if necessary list 2 substances?

I would like to take up the point - as the user of the one
truly accumulating site in the UK - that there is certainly
a strong argument that it is more desirable to contain
deposits in a limited sacrificial area than to scatter the
materials thinly over a much wider area and not know its
eventual fate. Are the Authors challenging the 'let's kick
it around until it gets lost' theory, or making a virtue out
of necessity?

MR W. HALCROW, Forth River Purification Board
The Authors of Papers 8 and 9 have given a reassuring
picture of UK practices but they have also drawn attention
to inherent uncertainty as to the safety of the process -
particularly in relation to far-field effects. Mr Harper
also advises against complacency. I would like to endorse
these three points and their importance in relation to
maintaining sea disposal as a long-term option for the UK.

The Marine Pollution Monitoring Management Group has
carried out a critical review of the working of UK disposal
grounds. This review paints a less reassuring picture of
our monitoring efforts. The principal conclusions of the
review are that a more comprehensive monitoring programme is
required, and we must attempt to address the problem of
potential far-field effects - possibly through the use of
the techniques used on the Thames ground. In essence our
efforts must be devoted to gaining public confidence and
obtaining international acceptance if we are to retain this
option.

The recommendations of the review have been accepted by
the DoE and MAFF and will shortly be submitted to the
disposal authorities. I would simply like to commend these
recommendations to the disposal authorities for action. The
costs of implementation will be trivial compared with the
benefits foregone if the industry loses this disposal
option. I would also like to hear the Authors' views on
whether or not more intensive monitoring is required and if
so who should pay for it.

MR D. HARRINGTON, WRc Processes
With reference to the question about digestion in Paper 9
(para. 12), digestion offers five advantages

    (a)   it reduces smell and improves handling
    (b)   it reduces pathogens by 90%
    (c)   it reduces total solids
    (d)   it may aid thickening characteristics
    (e)   it produces gas.

But if digestion is for pathogen reduction, as indicated
in the Paper, then 90% reduction is totally insignificant.
Have the Authors considered heat treatment, chemical or
storage? Digestion for the reduction of coliforms implies
lack of confidence in disposal routes. If digestion is an
aid to thickening then it is inefficient. Do the Authors
have any comments?

Finally, odour control is best served and more economic by
the other options, e.g. ducting the gases to compost.

MR J. A. WAKEFIELD, Coastal Anti-Pollution League
Would Mr Harper comment on the complaint from members of the
CAPL that algal bloom derives from sludge dumping in
Liverpool Bay?

DR D. LAU, Environmental Protection Department, Hong Kong
On Paper 9, what criteria were adopted for the selection of
the accumulative dumping site at Garroch Head?

If it becomes necessary for Strathclyde to search for
another new marine dumping site, would consideration still
be given to selecting an accumulative site, bearing in mind
the operational experience obtained from sludge dumpings at
both dispersive and accumulative sites in the UK?

MRS L. EVISON, University of Newcastle upon Tyne
Colwell (ref. 1) emphasized the possibilities for genetic
transfer among bacteria in sewage polluted marine waters,
and also the evidence for some pathogenic bacteria entering
a state of dormancy (although retaining virulence), and put
forward the view that ocean waste disposal should be
rethought.

As a microbiologist I believe these factors certainly
ought to be taken into consideration in determining policies
for disposal of sewage and sludge to sea, although I
certainly do not subscribe to the view that no sewage or
sludge should be discharged to the sea. The state of
dormancy of pathogens in sea water which has been reported
may last for a few weeks, but ultimately the pathogens will
become non-viable. To initiate infection in a human host,
an infectious dose of more than 1000 organisms for most
bacterial pathogens is required. Provided that sea outfalls
are designed so that adequate dilution and dispersion of
sewage occurs, low concentrations of the dormant bacteria
should not pose a significant health hazard to bathers. In
my view, since bacterial death in sea water cannot
apparently be relied on, design calculations should not
incorporate T_{90} values.

As a further safeguard I would suggest that it would be
advisable to design outfalls so that bathing water complies
with the EEC guideline rather than the mandatory bacterial
standards, since these standards are more similar to those
recommended by the EPA, based on a sound epidemiological
study.

If unfavourable hydrographic conditions exist so that a long outfall cannot ensure compliance with the guideline standards (faecal coliforms 100/100 ml in 80% samples), then some form of disinfection, for example by chlorination or the Clariflow system with a lime-based coagulant, could be used.

PROFESSOR McINTYRE and DR PARKER, Paper 8

In reply to Mr Brown, if it is not possible to find an acceptable disposal site in coastal waters, then shipping sludge to the open ocean is worth attention. The two major considerations are financial and environmental. On the first point, if the nearest deep water is far from the coast and a long sea journey is necessary, then the cost could become a very significant factor. On the second point, the method of disposal would need to be carefully worked out. Discharge at or near the surface could cause problems if the material were carried back to the land. The optimum approach would be to discharge as near as possible to the sea bed. Given reduced water movements at great depths, the sludge from each operation would tend to accumulate on the bottom, and since the deep water benthos has evolved in relatively stable conditions, the sudden introduction of a nutrient load changing the characteristics of the sediments would certainly have an impact on the fauna. Some objectors would no doubt wish to keep the deep sea pristine, but it could be argued that since only a minute fraction of the sea bed is affected, this is a small price to pay for the use of the oceans. It is worth noting that the operation would provide a valuable opportunity to observe the reactions of the benthos.

In reply to Mr Halcrow, monitoring as carried out at Garroch Head and from Edinburgh which is done to the requirements of the licensing authority and paid for by the disposer, is in my view entirely adequate at present.

MR HARPER and MR GREER, Paper 9

Mr Brown is correct that the Mersey Valley sludge pipeline is about to come into operation. The sludge that arises from inland towns such as Bury, Bolton, etc, at present goes to sea after being transferred by tanker vehicle to Manchester Davyhulme sewage treatment works and hence by boat along the ship canal. The move to Liverpool for the inland sludges will basically only affect the shipping operation and not give rise to any further sludges. There will be some increase, as to minimize the environmental impact of tanker vehicle movement there has been some storage in lagoons pending the pipeline completion, and this will be phased out. The new sludges which we plan to ship to sea will arise from first time sewage treatment for towns draining to the Mersey estuary. Even primary treatment produces a large quantity of sludge and we estimate Merseyside will produce up to 20 000 dry tonnes per year in the next decade or so as crude sewage is removed from the estuary and treated.

The NWWA, like all water authorities, have vigorous trade effluent control policies for discharges of list 1 and list 2 substances to both the river and to sewers. The Paper illustrates the success for mercury and there are similar approaches to all list 1 substances.

In reply to Mr Halcrow, although I believe that sludge to sea properly controlled is a good disposal route and can represent the best practicable environmental option, there are still uncertainties. The best route should be selected based on a good assessment and understanding of the risks involved, so I believe we should increase our understanding of the sludge to sea route and that means more extensive and intensive monitoring. I believe this statement also applies to other routes. There should be a contribution nationally rather than just locally, but I also accept that it is inevitable that water authorities - and this means the local community - will have to fund much of the work.

In reply to Mr Wakefield, the origin of algal blooms in Liverpool Bay has been studied for many years and the exact reasons for them are still unknown. I believe it is a fair summary of scientific opinion that there is no evidence to suggest that sewage sludge makes a significant contribution to algal blooms.

REFERENCE

1. Colwell R. Microbiological effects of ocean pollution. International conference on environmental protection of the North Sea, London, 1987.

10. A sewage treatment strategy for the Colwyn and Aberconwy areas — an appraisal of options

C. PATTINSON, BSc, PhD, Principal Scientific Officer (Tidal Waters), and R. JONES, BSc(Eng), MICE, Project Engineer, Welsh Water Authority

SYNOPSIS. Considerable problems attach to the formulation of capital investment strategies for sewage treatment in coastal areas due to the difficulties involved in obtaining reliable flow information and in accurately estimating the cost of sea outfalls. Difficulties may also occur due to the necessity of incorporating intensive expenditure on outfall contracts within a limited Capital Programme. This case-study describes how these problems were overcome by developing a set of standard cost models, by carrying out sensitivity analyses and by adopting specific targets for phasing of capital expenditure.

INTRODUCTION

1. The Conwy Estuary, Llandudno and Colwyn Bay together form one of the most popular tourist areas in the U.K., there being in excess of 50,000 holiday bedspaces in an area with a resident population little greater than 60,000 (Fig 1). The estuary and coastal waters have a very high amenity value and it is estimated that over 30% of all visitors engage in some form of water – based recreation, much of it based on the beaches at Llandudno North and West shore and at Deganwy and Conwy Morfa. The outer estuary and Conwy Harbour are also very popular for sailing and windsurfing whilst the Harbour area is a tourist attraction in its own right. In addition, the coastal area supports a thriving commercial and sport fishery whilst the estuary is one of the most important migratory fisheries in Wales. There is also a significant commercial shellfishery in the mouth of the estuary.

2. The coastal waters receive considerable quantities of sewage effluent through a total of 13 sewage outfalls, most of these being old, poorly sited and providing little or no treatment. This has resulted in significant aesthetic and bacteriological pollution.

3. For a number of years investigations and studies have taken place including both on-shore and off-shore surveys into suitable locations for outfall and sewage treatment works discharges.These investigations culminated in the adoption, in 1979, of a strategy for improving the

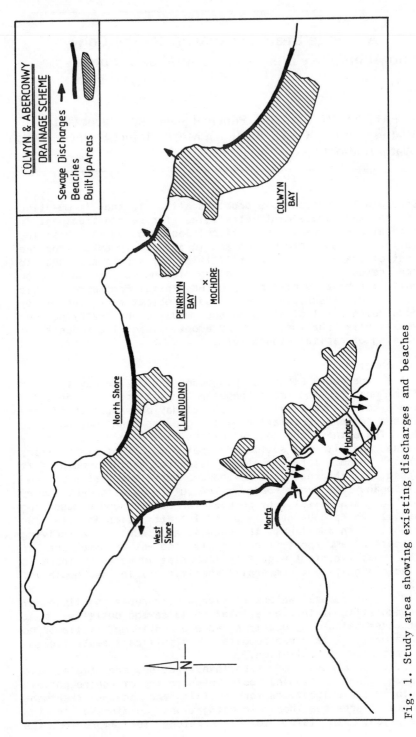

Fig. 1. Study area showing existing discharges and beaches

situation. This strategy however could not be implemented as it required considerable capital investment over a short period of time. This proved impossible to accommodate within the Authority's Capital Expenditure Programme due to the competing demands of other high priority schemes.

4.   Following its 1984 reorganisation the Authority determined to review the situation with a view to formulating an acceptable capital investment strategy.

EXISTING SITUATION

5.   The Authority's existing sewage disposal assets in the area give rise to significant environmental and engineering problems. They are generally aged, being in the region of 60 − 100 years old with the exception of the holding tanks and headworks for the main West Shore, Llandudno outfall, which are some 20 years old. Despite their age, they have caused few maintenance problems with the exception of occasional protection and replacement   of the seaward extremities of the outfalls. Their future design life however (particularly the cast−iron outfalls) is considered to be short−term only.

6.   From an environmental perspective, of the 13 discharges, 11 fail to meet the Authority's own minimum level of service for discharges to tidal waters. discharging at or above low water and giving rise to aesthetic contamination with sewage debris on the adjacent shoreline, resulting in significant and sustained public complaint. Of particular concern in this respect are those outfalls directly affecting high amenity areas such as Conwy Harbour and the beaches at Conwy Morfa, Deganwy and Llandudno West Shore.

7.   All the outfalls in the area also contribute to the generally elevated bacterial levels found in the water and whilst these elevated bacterial concentrations do not in themselves constitute a serious risk to public health, they do fail to meet the standards specified in the EEC Bathing Water Directive, a standard increasingly identified by the public with safe and acceptable bathing water quality. Conversely, the Llandudno North Shore beach is of excellent quality, easily achieving the EEC standard and whilst the coastline in the vicinity of Penrhyn Bay is subject to aesthetic contamination, this area is of limited public accessibility and   recreational value and is thus of less concern. Colwyn Bay, also occasionally fails to comply with EEC standards but here the problem is primarily attributable to storm water overflows and can be overcome in the short term by sewerage improvements.

8.   The elevated bacterial levels also contribute to contamination of shellfish beds which are subject to a closure order which requires that the shellfish are cleansed prior to sale. This is carried out at the Conwy Cleaning Station which abstracts water from the middle reaches of the estuary and which pretreats and sterilizes the water prior

to use.   The sea fishery is unaffected by the discharges which also have no affect on the viability of the migratory fishery, the estuary remaining well oxygenated and with low ambient ammonia levels at all tidal states.

CAPITAL INVESTMENT OBJECTIVES
9.   The initial studies indicated that the Authority's minimum level of service was not being maintained and that the   future   asset   life of the existing outfalls was restricted.   This, together with the increased public awareness of problems associated with beach pollution dictated that further studies should be carried out, whereby various options for improving the situation would be appraised and a strategy for Capital Investment formulated.
10.   The major objectives of the further studies were to formulate a Capital Investment Strategy whereby:-
a) The Authority's ultimate level of service for tidal waters should be achieved within a reasonable time period.
b) Capital Investment should be phased so that the discharges causing the greatest pollution should be dealt with first.
c) Capital Investment, when considered together with the resultant effect on revenue costs, should provide the most cost effective investment necessary to comply with (a).
d) The Capital Investment required should be capable of being phased so as to minimise the effect on the remainder of the Division's Capital Programme.
e) The choice of strategy would be unaffected by any future detailed investigations.
f) The strategy should be capable of implementation in harmony with the concurrent major road-works.

FACTORS AFFECTING THE STRATEGY
11. Strategy formulation for large-scale sewage treatment problems is extremely difficult due to the many unknown factors involved.   In particular it is extremely difficult to accurately estimate the costs of sea outfalls and headworks and unless comprehensive flow monitoring has been undertaken over a long period, flow information will be far from reliable.   The philosophy behind the method used to formulate this strategy was to determine a set of models and procedures whereby all the items which could be considered to be significant in terms of cost were dealt with on a common basis. This method enabled a large number of options to be considered over a relatively short time period and led to the selection of a strategy which could be adopted with confidence as being the most cost-effective in all likely future circumstances.
Design Criteria
12.   The Authority's requirements with regard to environmental standards are that if significant capital is to be invested, then the improvements should be designed to the Authority's ultimate or long-term level of service.

With respect to this strategy this meant that any remedial scheme must assure that all the legitimate uses of the estuary as indicated in the introduction to this paper should continue to be protected. In order to do this, environmental quality standards were specified for critical water quality parameters to protect the public engaged in bathing and other water contact sport and to preserve the migratory fishery. In view of the high amenity value of the area it was felt appropriate that any scheme should achieve the EEC bathing water bacterial standard on the LLandudno, Deganwy and Conwy Morfa beaches and also in the Conwy Harbour area; this standard would only be applied at the waters edge on recognised beaches and not on rocky or otherwise inaccessible areas of the coastline. In order to protect the passage of migratory fish an oxygen standard of 5 mgl−1 was specified. Normal practice is to specify an standard of 1 mgl−1 for ammonia to protect the passage of fish but it was felt that this would result in an unacceptable risk of excessive macro−algal growth in the middle reaches, of the Conwy and accordingly a much more stringent standard of 0.3 mgl−1 was specified to inhibit such growth.

Capital costs

13. The items considered significant from a capital point of view were outfalls, headworks, treatment works, pumping stations and pipelines. All the options considered contained at least some of those items, so consideration was given to the best way of costing these items to give a common approach over all the options. No problems are generally experienced in costing pipelines and pumping stations due to the large amount of data available. Reliable information on outfalls, headworks and treatment works is, however, usually sparse and to overcome this a data collection exercise was undertaken with the aim of collecting sufficient information to formulate empirical functions, which would quickly give estimated capital costs for these items at a fixed price level date.

14. Outfalls:− Over recent years, estimating the cost of outfalls has proven notoriously difficult. This is no doubt due in part to the dissimilarity of the conditions applying to those outfalls which have been constructed and also due to the relatively few outfall schemes which have been implemented. It does appear however that the trend in outfall costs has been steadily downwards due in part, perhaps to spin−offs from off−shore oil industry technology but also because of increased competition due to more contractors moving into the marine sector.

15. In order to obtain as much reliable information as possible, a telephone survey was carried out of all Water Authorities and Scottish Regional Councils to ascertain details of location, length, diameter and cost of outfalls which had actually been constructed and also contractor's budget quotes for outfalls soon to constructed. The results

of this survey are shown in Fig 2.   An attempt was made to
correlate the results in a function which measured
cost/length against diameter with an appropriate allowance
for setting up site costs.   A single function which gave a
seemingly reliable comparison with actual costs was
determined and is also shown in Fig 2.

WORKS COST(£) = DIAMETER (mm) x (LENGTH (m) + 500)

    This function was used to estimate works costs for all
outfalls considered in formulating the strategy.

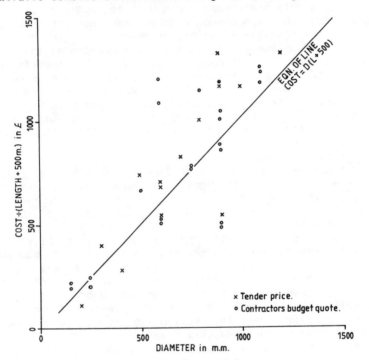

Fig. 2. Outfall costing model

    16.   Headworks:- A similar telephone survey procedure was
followed to determine costs for headworks structures.   More
difficulty was experienced here as each headworks proved to
be completely individual with total costs dependent on
siting of headworks, whether it was buried or exposed, the
degree of pretreatment and on the particular requirements of
each   Authority concerned.   Particular difficulty was
encountered when considering the level of pretreatment and
the capital/revenue cost implications of removing non-
biodegradable solids from the effluent should this prove
necessary.   In the event, the cost function evolved included

for traditional pre-treatment of screening, maceration and grid removal only, due to the lack of reliable information on any enhanced level of pretreatment. The function chosen for this strategy after analysis of all the survey data received was:-

WORKS COST = 1000 X DIAMETER OF OUTFALL (mm)

17. Sewage Treatment works :- The telephone survey approach adopted previously was not feasible, there being few recent examples of new S.T.W.'s constructed on green field sites which were not "special cases". In order to formulate a function for sewage treatment works costs a combination of T.R. 61 costs and costs obtained from the STOM programme was used. A treatment works was assumed consisting of preliminary treatment, primary sedimentation, aeration, secondary sedimentation and sludge digestion together with interconnecting pipework and siteworks. STOM was to estimate costs for the main process units with TR61 for the remainder. It was also necessary to build into the function an allowance for the fact that both S.T.W. sites available were situated on very poor ground requiring ground consolidation at an estimated cost of £250,000 /ha. The functions determined and used in costing the options were:-

S.T.W. with secondary treatment :- WORKS COST (£m) = 0.92 + 0.32 D.W.F. (MI/d).
S.T.W. with primary treatment :- WORKS COST (£m) = 0.79 + 0.15 DWF (MI/d).

Revenue Costs

These were investigated in two ways:-

18 Any of the options examined would incur additional revenue costs in the form of power, supplies and staff. Generally speaking each of the options would involve pumping a similar volume of sewage through a similar head with the same number of pumping stations. It was therefore decided that because differences in revenue costs for pumping were unlikely to be significant between options in terms of the overall strategy, comparisons in revenue costs would be confined to the extra revenue costs of sewage treatment works options over those options which did not include S.T.W.'s

19 Costs for operating sewage treatment works were based on historical data already held by the Division and on the STOM model. The function determined is in terms of N.P.V. of revenue costs over 20 years and includes an allowance for increase in flows of 1.5% per annum:-

N.P.V.(20 yrs) of revenue costs for Primary Sewage Treatment Works is £15,000 Q
N.P.V. (20 yrs) of revenue costs for Secondary Sewage Treatment Works is £25, 000 Q
Q = present D.W.F. is I/s

Phasing — Economic

20. The strategy which had been produced for this area in 1979 had been capital intensive, calling for large sums of capital to be invested over a short period. One of the difficulties of implementing this strategy would have been to inhibit capital being expended elsewhere on other high priority schemes and this was the major reason that the strategy was not implemented. In order that the same constraint should not apply to the new policy it was therefore imperative to investigate the possibilities of phasing the work inherent in each option investigated. On examination of the projected capital availability over suceeding years a target figure of £3m was established as the maximum which would be available per annumn without inhibiting other high priority capital investment in other areas.

Phasing — Environmental

21. The priority areas for improvement from an environmental viewpoint were the Conwy/Deganwy/Llandudno Junction areas, followed closely by West Shore, Llandudno. The situation at Penrhyn Bay and Colwyn Bay was of a lower priority. Ideally, any phasing of the work should enable the high priority areas to be tackled first.

Flow Predictions

22. As is general with studies of this type, flow information was limited and the data available was only considered to be useful as background information and not as the basis for formulating a strategy. It was therefore necessary to devise some means whereby the sensitivity to flow changes of the options examined could be compared, in order to establish whether the choice of option could be affected if the physical flow data differed from that estimated.

23. Each option considered was therefore costed (using the functions already described) over a whole range of flows between an upper and a lower confidence limit. The upper limit was based on high per capita flows and a generous increase from estimated present day flows to 2021 flows. The lower limit was based on low per capita flows with no increase from present-day flows. A "best-estimate" figure for future flows was established using the data which was available. This figure fell within the two confidence limits and was the figure on which the proposals for future capital investment was based. The variance of cost with flow for each of the options considered is shown in fig 5.

Roadworks.

24. Major roadworks will be taking place in the area over the next few years in connection with the A55 improvements, including a tunnel beneath the Conwy estuary. It was therefore necessary to ensure that the proposed strategy could be phased so as to ensure that the respective construction works did not clash and that the construction of the Authority's scheme would not add to the considerable

traffic congestion already expected from the road construction works.

Economic Analysis

25. In normal circumstances the capital investment strategy for a scheme of this extent would be based on a comprehensive economic analysis taking into account all aspects of capital costs, revenue costs, replacement costs and residual values over a 60 year period. The most cost effective investment for the Authority would then be determined by a comparison between the various options, examined on a Net Present Value basis.

26. The time available to complete the investigation and the lack of detailed information mitigated against this approach, so it was decided to compare options on an N.P.V. basis over a 20 year period taking into account capital and extra over revenue costs for sewage works but not considering replacement costs or residual values. Depending on the capability of each option to be phased accordingly, the N.P.V.'s of the capital costs were based, where possible, on the following phasing which follows the environmental priorities:-

    i) Conwy, Deganwy, Llandudno Junction – No delay

    ii) Llandudno –+ 5 years

    iii) Colwyn Bay, Penrhyn Bay –+ 10 years & + 20 years

Two alternative phasings were examined for Colwyn Bay/Penrhyn Bay.

CONSIDERATION OF SOLUTIONS
General

27. The options available for a comprehensive sewerage and sewage disposal scheme in the area are restricted by many environmental and physical constraints. All schemes would comprise the transport of sewage flows via a system of pumping stations and rising mains to the ultimate treatment site(s) which would comprise a long sea outfall or a sewage treatment works or a combination of both. As the pumping stations and rising mains would be similar for each option, the main differences between the options was the choice between, and the siting of, the main treatment units. The options selected are described in fig 3 and are all capable of achieving the desired quality standards.

Long Sea Outfalls

28. To the west of the area the locations for a long sea outfall were restricted by the proximity of the mouth of the Conwy Estuary with its complex flow patterns and extensive inshore shallows. Physical constraints including the scarcity of suitable sites for launching an outfall added to the siting problems. Only one site was identified as being suitable for long sea outfall discharges, in an area some considerable distance to the west of the existing Llandudno West Shore Outfall. Two alternative launching sites were available, one at West Shore and one at Conwy Morfa.

29. An alternative location was identified to the N.E. of Penrhyn bay where a suitable launching site is available. No locations off Llandudno N.Shore were considered, as this is by far the most popular beach in the area, complies with E.E.C. conditions and is free from any current foul sewage discharges.

30. The performace of options in the Llandudno West Shore and Penrhyn Bay areas in relation to the quality standards was investigated using conventional hydrographic techniques to describe initial and secondary dilution. These various procedures determined the optimal lengths and levels of treatment required to achieve the desired quality standards in the receiving waters for a wide range of potential flows.

Sewage Treatment Works

31. The major constraints for options containing a sewage treatment works were the environmental and planning issues raised by the siting of a treatment works in an environmentally sensitive area. Only two sites were identified as possible locations, one in the south of the area close to Llandudno Junction and one in the east close to Penrhyn Bay at Mochdre.

32. Discharge conditions from the Mochdre site were determined by using the information available from the Penrhyn Bay Hydrographic studies. This indicated that the treatment necessary would be primary settlement only as long as the flow was discharged via a medium length outfall.

33. For the Llandudno Junction site discharge would be to the Conwy Estuary and it was therefore necessary to carry out investigations into the location of discharge and level of treatment required. Evaluation of the effects of various estuarine discharges on water quality in the area was undertaken by using 1 - d steady state moving segment model specially developed for this purpose. The model was used to predict the performance of the options for a variety of sewage flows and levels of treatment in relation to the quality standards for ammonia, oxygen and bacteria. The near field effects of a number of potential outfall sites within the estuary were determined by dye releases. These investigations indicated that the optimum position for discharge was in the centre of the channel some 1km from the treatment works site.

Development of Options

34. After considerable evaluation of the possibilities, nine options were selected for detailed investigation and costing. These options covered the whole range of treatment options and sites available, some options dealing with the problem by means of two or three separate phased schemes, others by phased development of one combined scheme (Fig 3).

EVALUATION OF OPTIONS

Costs

35. Overall costs of the options examined are shown in fig 4. This indicates that the least expensive option in

Phase / Option	Phase I — For flows from Llandudno Junction Deganwy and Conwy	Phase II — For flows from Llandudno	Phase III — For flows from Colwyn Bay
1	Sewerage and pumping stations. Sewage treatment works at Llandudno Junction.	Outfall at West Shore	Sewerage Outfall at Penrhyn Bay
2	Sewerage and pumping stations Sewage treatment works at Mochdre	Outfall at West Shore	Sewerage Extensions to S.T.W.
3	Combined Phases I and II – Outfall at West Shore – sewerage and pumping stations to transfer flows from Llandudno Junction/Deganwy/Conwy		Sewerage Outfall at Penrhyn Bay
4	Sewerage and pumping stations Outfall at Penrhyn Bay	Sewerage and pumping station to transfer flow to outfall	Sewerage Outfall at Penrhyn Bay
5	Sewerage and pumping stations Outfall at Penrhyn Bay	"	Sewerage
6	Sewerage and pumping stations Sewage treatment works at Llandudno Junction	Sewerage and pumping stations to transfer flow to S.T.W.	Sewerage Outfall at Penrhyn Bay
7	Sewerage and pumping stations Sewage treatment works at Mochdre	Sewerage and pumping stations to transfer flow to S.T.W.	Sewerage Extensions to S.T.W.
8	Sewerage and pumping stations Outfall at Mochdre	Outfall at West Shore	Sewerage Outfall at Penrhyn Bay
9	Sewerage and pumping stations Outfall at Morfa	Sewerage and pumping stations to transfer flow to outfall	."

Fig. 3. Options appraised

169

terms of capital and capitalised running costs is Option 5, (a long sea outfall at Penrhyn Bay for the whole area) followed by Option 3 (an outfall at West Shore and a separate outfall at Penrhyn Bay).

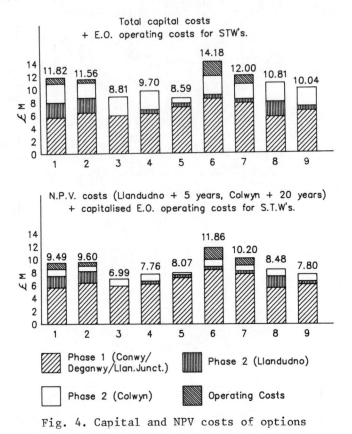

Fig. 4. Capital and NPV costs of options

36.   When N.P.V. costs are considered however, the fact that the outfall at Penrhyn Bay could be deferred in Option 3 because of its low priority, makes Option 3 the most economic option.   Indeed if the solution of the problem of Colwyn Bay could be deferred by 20 years then Option 4 becomes more economic than Option 3.

Flow Sensitivity

37.   The effect of changes in flow are shown in fig. 5.

This shows the relative N.P.V. costs of the options with Llandudno delayed 5 years and Colwyn delayed 20 years (where the options allow such phasing).   From this it can be seen that the cost of the options is not significantly flow sensitive , Options 3, 4, 5 and 9 being considerably more economic than all the other options for all flows, with Option 3 being noticeably the most economic for this phasing.

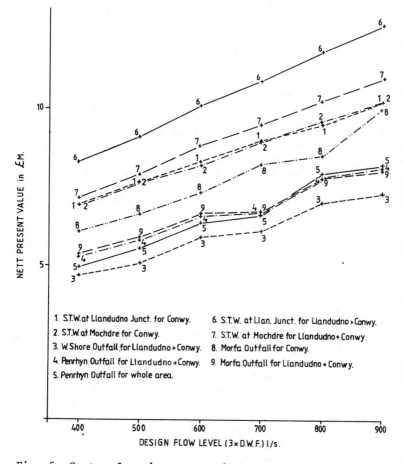

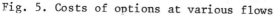

Fig. 5. Costs of options at various flows

Type of Treatment

38.   Fig. 5 indicates that in all circumstances all the
options containing outfalls (Options 3, 4, 5, 8 and 9) are
more economic than all the options containing treatment
works, even allowing for the recognised uncertainties
involved in costing outfalls.

Phasing

39.   The least cost capital solution (Option 5) would
involve the construction of a long sea outfall at Penrhyn
Bay as the first phase, pumping stations and rising mains
from Conwy, Deganwy and Llandudno Junction as the second
phase and the connection of Llandudno as the third phase.
This phasing is unfortunate in that it requires expenditure
of the whole of the capital sum over a period of 5 years,
whereas it is not essential to construct the solution to the
Colwyn problem for some 10 – 20 years due to the lack of
urgency on environmental grounds.

40.   The most economic solution overall, bearing in mind that it allows for the later delay of the Colwyn solution is Option 3. This would involve the construction of a long sea outfall at West Shore as phase one, pumping stations and rising mains from Conwy, Deganwy and Llandudno as the second phase and the construction of an independent outfall at Penrhyn Bay at a later date, to be determined. Although not fitting precisely with the environmental priorities, (as Llandudno is solved before Conwy/Deganwy/Llandudno Junction), the Option complies more closely than Option 5 and does allow for phasing of capital over a number of years.

41.   When considering the effects of the A55 road construction it is important that any work in Llandudno Junction/Deganwy (Where the major clashes occur) should not be undertaken in the early years of the strategy. It is also important that no work should be carried out in the narrow, winding streets of Conwy until the road tunnel by-passing the town is open. Both Option 3 and Option 5 comply with these requirements although many of the other options do not.

SELECTION OF STRATEGY.

42.   The most appropriate Option for adoption as a capital investment strategy was considered to be Option 3 - a long sea outfall at West Shore, followed by linking it with Conwy/Deganwy/Llandudno Junction flows, followed some time later by a long sea outfall at Penrhyn Bay.

43.   This option was the most economic in N.P.V. terms (assuming a delay of the Penrhyn Bay outfall), allowed for phasing over a number of years (thus easing the annual capital demand), and was acceptable in complying with the environmental and roadworks constraints. No other matters considered, including the possible change of flows were significant enough to affect this conclusion.

44.   The strategy has now been included in the Capital Programme for commencement on site in 1989.

CONCLUSION

45. The approach to option appraisal described in this case study enabled a strategy to be adopted with confidence in a limited time period,without the need to carry out detailed investigations into flow figures and costings.

46.   The results of the option appraisal indicated that for this area all options containing long sea outfalls were significantly more economic than all options containing treatment works.

47.   The strategy containing the selected option has now been adopted and incorporated into the Authority's capital Programme with the first phase due to commence in 1989.

11.   Thanet sewage disposal investigation and the comparison of joint and local solutions

F. N. MIDMER, Director of Technical Services, and
M. J. BROWN, Project Engineer, Southern Water Authority

SYNOPSIS.   The Isle of Thanet is an important holiday area in East Kent with a total summer population of 180,000 in the four seaside towns of Margate, Ramsgate, Broadstairs and Birchington.   Sewage is discharged to sea through short sea outfalls causing pollution of local beaches.   The need for improvements, which had been discussed for many years, was finally brought to a head when Margate was designated a bathing water in accordance with EC Directive 76/160/EEC.   The paper outlines the options examined, which varied from a single outfall for all the towns, through separate schemes, to a single treatment works and the consideration given to the resource benefits of effluent re-use.

BACKGROUND
   1.   The sewerage systems of the Thanet towns of Birchington, Margate, Broadstairs and Ramsgate are largely combined and based on the major works carried out at the turn of the century.   Sewage is disposed to sea through a number of short sea outfalls resulting in significant pollution of local beaches.
   2.   Routine monitoring by the Southern Water Authority has shown that EC requirements for bathing waters are not met over much of the coastline.   The situation has long been considered unsatisfactory by the general public and, due to the importance of tourism in Thanet, by the business community.
   3.   The introduction of Directive 76/160/EEC (ref. 1) and the subsequent designation of Margate main beach as a bathing water enhanced the priority for action on the existing sea outfall problem.
   4.   The Authority considered that the adjoining towns of Birchington, Broadstairs and Ramsgate should be included in a full investigation of sewage disposal in the Thanet area allowing appraisal of local and area solutions.   Due to the geographical proximity of Sandwich, with similar problems of inadequate sewage disposal, this town was also included in the investigation.   The village of Minster in Thanet with existing satisfactory treatment works was included for completeness. The area of investigation is shown in Fig. 1.   Summer populations and flows are included in Table 1.

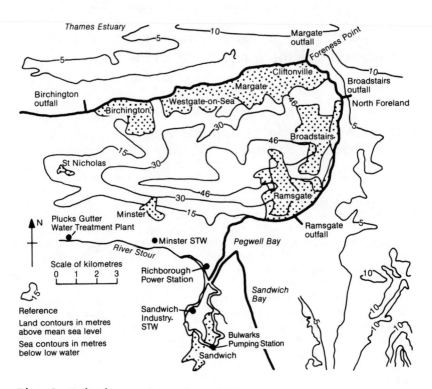

Fig. 1. Existing works and outfalls

Table 1. Summer population and flows 2001

Town	Population	Flow:m^3/d
Birchington	12630	3330
Margate	72080	21900
Broadstairs	37010	11220
Ramsgate	62750	19470
Minster	6710	1920
Sandwich Industry	–	5680
Sandwich Town	7870	2260

5. Following an inquiry into the Broad Oak Reservoir promotion in east Kent, the sewage treatment investigation was broadened to include consideration of the resource value of any effluent from Thanet, as it was thought that it might be a cost effective way of meeting the need for water resources.

SCOPE OF THE INVESTIGATION
6. At the time the investigation was undertaken there was a general presumption within the Water Authority (as borne out by earlier studies), that sea disposal was the most cost effective method of achieving satisfactory sewage disposal for coastal towns. There was, however, general public opposition to sea outfalls. It was considered essential therefore to carry out a thorough examination of a wide range of options, ranging from individual sea outfalls or inland treatment works for each town to a single area treatment works or sea outfall solution and including combinations of options between these two extremes.
7. The work divides naturally under five headings: Sea Treatment; Works Treatment; Sewerage Transfer System; Resource Benefits and Economic Appraisal, and these headings are used for convenience in the paper. Details of design standards are included in Appendix 1.

SEA TREATMENT
8. Marine consultants carried out a wide-ranging survey of the north east Kent coastline, concentrating on possible discharge points which could be sensibly reached from the existing outfall headworks positions.
9. The sea surveys included hydrographic, float tracking, dye and bacteriological tracer surveys conducted at a number of tidal states and throughout a neap/spring tidal cycle to ensure that the many variations in tidal water movement were observed. Long term wind records were obtained from the Meteorologial Office for analysis. Geophysical surveys of the sea bed were also carried out.
10. These surveys showed that the North East Kent coastline is generally favourable for sea disposal. There are fast tidal currents providing good dilution and dispersion and a general residual movement of water to the north-east away from the coast into the North Sea. In addition, the sea bed is either sand or chalk rock, neither of which will present undue problems in dredging and burying an outfall. The disadvantage, however, is that the coastal waters are relatively shallow and water depths necessary are only found at distances in excess of 1500 metres from the shore.
11. Suitable sites which complied with Authority standards were identified at Foreness, North Foreland and Ramsgate. These are shown on Fig. 2 and briefly described below.

Foreness
12. The discharge site lies some 1900 metres offshore and to the immediate north of the existing outfall station, in 20 m

of water. The headworks would be adjacent to the existing outfall station and similarly constructed in the chalk cliffs and below the effective skyline. The outfall construction area would be a difficult cliff top site some 15 m above the foreshore.

North Foreland

13. A discharge site lies some 3200 metres offshore and to the east of the existing Broadstairs Outfall Station in 11 metres of water. The headworks would be adjacent to the existing outfall station and could similarly be constructed in the chalk cliff and below the effective skyline.

14. The land inshore would provide a suitable construction area. The disadvantage is that Joss Bay is used extensively throughout the summer months and the contractors use of the beach area would need to be restricted.

15. This site would be suitable either for the treatment of the present flow or as a regional outfall to which other flows would be pumped.

Ramsgate

16. A suitable discharge site lies some 5000 metres offshore and to the south-east of the existing Ramsgate Outfall Station in 11 metres of water. The headworks would be located within the existing dock area.

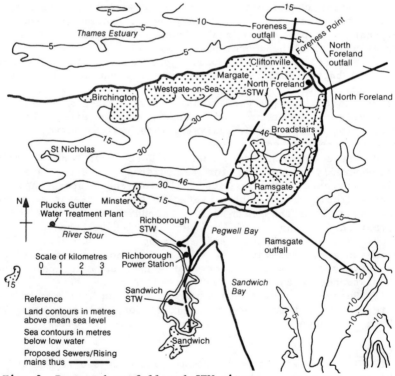

Fig. 2. Proposed outfall and STW sites

17. It would be difficult to construct and launch the outfall from the Ramsgate site, but a stringing yard would be temporarily reclaimed in the adjacent Pegwell Bay which has the disadvantage of being a Site of Special Scientific Interest.

WORKS TREATMENT

18. A number of sites were considered for the location of treatment works, both inland sites and coastal sites discharging to the sea. It was considered that coastal treatment would avoid the transfer costs associated with inland treatment and thus reduce the expected cost differences between inland treatment and sea disposal. This option was therefore examined in detail but for the reasons discussed below did not prove satisfactory.

19. Discharges to the sea were required to achieve EC standards for bathing water, including the standard for bacteria reduction. It is probably this standard that is the most fundamental factor in achieving satisfactory sewage disposal from coastal towns.

20. A long sea outfall, properly sited and properly designed, will achieve the bacteria reductions required by dilution and die-off in the body of the sea.

21. Conventional sewage treatment, however, achieves only a marginal reduction in bacterial concentrations and a treated effluent cannot be discharged directly to the sea adjacent to a bathing water. Further treatment is required perhaps by discharge through a long sea outfall or discharge inland to a watercourse which has a sufficiently long journey time to the sea for die-off of bacteria to occur. Disinfection by chlorination is feasible but the Ministry of Agriculture, Fisheries and Food have advised that the organochlorine compounds formed by chlorination (ref. 2) would not be acceptable to fishery interests. All these factors add to the cost of conventional sewage treatment.

22. In addition, coastal sites in an area such as Thanet have high amenity value and sewage installations require careful siting and treatment to minimise visual intrusion, noise and smell. Some degree of disinfection of ventilated airs to control odours, particularly when sludge is being handled, is required. These problems apply similarly to outfall headworks but these installations are smaller and generate less odour.

23. In the event there were only two serious options for the siting of a major treatment works; one at North Foreland, discharging to the sea, and the other at Richborough, discharging to the tidal section of the River Stour. There was insufficient land in the Ramsgate area and although there was sufficient land at Foreness to serve Margate the land has high amenity value and the works was required to be constructed below ground level and fully enclosed. A scheme was developed but with a NPV of £13.8M was unacceptably costly. The existing site at Minster was considered but the

site is more expensive than Richborough in terms of development and sewage transfer. A suitable site for a works serving Sandwich only was located north of the town with discharge to the tidal Stour. The sites are shown on Fig. 2 and described in paragraphs 28 to 30.

Effluent Standards – Sea Discharge

24. For a coastal treatment works discharging directly to the sea it was considered that the principal objective should be solids removal. The proposed effluent standard was SS 150 and this would be achieved with the provision of primary treatment. It was considered that a more stringent standard including BOD reduction, and therefore involving greatly increased costs of secondary treatment, was not warranted; first, because of the natural abundance of oxygen in the sea at the discharge sites, secondly, because the effluent would in any case have to be discharged through a long sea outfall to achieve the required bacteriological standards on adjoining beaches and the bacteria reduction achieved by secondary treatment would not have a significant effect on the outfall length.

Effluent Standards – River Discharge

25. The tidal River Stour is an important part of the river system and is classified as Class A, (good quality). The Authority policy is to maintain this quality and to achieve standards for estuarial waters required by the European Inland Fisheries Advisory Commission. A large area works would, however, discharge effluent equivalent to approximately one third of the dry weather fresh water flow in the estuary and would clearly have a significant impact on river quality.

26. The Stour Estuary model, developed to assist in hydraulic and saline instrusion calculations in the tidal Stour, was able to study the transport, degradation and interaction of pollutants including BOD, nitrogen and nitrates. The model was used to examine the quality /environmental effects of various effluent standards.

27. The effluent standard required for a regional works serving Thanet was SS 30, BOD 20, Ammonia 5 mg/l, on a 95 percentile basis. Where the works served a smaller part of the area, such as Ramsgate, a relaxation in ammonia standard would be possible. For a small works serving the Sandwich area only, the standard required would be SS 60, BOD 40 mg/l, on a 95 percentile basis.

Richborough Site

28. The Richborough site is adjacent to the existing Richborough Power Station, has good access and is reasonably remote from housing. It would be possible for effluent from this site to be discharged either to the river upstream of Sandwich or to the river mouth, or directly to the sea, but the latter two options would not be acceptable from bacteriological considerations on the adjoining bathing beaches.

Sandwich Site

29. A small treatment works for the town and adjoining industrial area would be located on land adjacent to an existing privately owned treatment works. The site has poor access but is remote from housing and not subject to planning difficulties.

North Foreland Site

30. A suitable site exists on farmland to the rear of the existing Broadstairs outfall station. The site has good access and is remote from housing but is adjacent to Joss Bay which is extensively used throughout the summer months and the works would require careful siting and design to minimise visual intrusion, noise and smell.

SEWERAGE TRANSFER SYSTEM

31. A number of options considered involved the transfer of sewage from the existing discharge location to a new site. In the Isle of Thanet land is relatively high, particularly in the eastern part, and transfer by pumping main would involve high pumping heads. Two alternatives were therefore examined, first, transfer by pumping main with a new pumping station at the existing outfall location and, secondly, transfer by gravity tunnel deep under the high ground with pumping as necessary at the new collection site. Taking into account capital and running costs, transfer by pumping main was some 15% cheaper.

32. There is room at each existing outfall location to construct a new pumping station but these were found to be relatively expensive due to the considerations regarding amenity which have already been discussed. The running costs, mainly power, were found to be significant and for schemes requiring sewage transfer the costs of transfer became a significant item.

33. The routes of the trunk sewers are shown on Fig. 2.

34. At each existing outfall location consideration was given to the provision of improved storm overflow arrangements consisting of storm water storage, screening and, where necessary, additional pumping. The necessary structures would be incorporated with the sewage transfer pumping station where provided.

WATER RESOURCE BENEFITS OF EFFLUENT RE-USE

35. The investigation considered the value of the sewage effluent from Thanet as an addition to the water resources of the area following a public inquiry into the proposed Broad Oak reservoir.

36. The reservoir required increased abstractions from the River Stour and the opposition considered that, first, the reservoir could be avoided if Thanet effluent was discharged to the Stour for subsequent abstraction and treatment. Secondly, reduction in flow resulting from increased abstractions would allow salt water to move further upstream

affecting abstractions by farmers for spray irrigation and by Richborough Power Station.

37. After careful consideration it was decided that the most cost-effective method of effluent re-use would be discharge of a treated effluent at the Richborough site, close to the summer upstream salt water/freshwater interface, to replace an equal quantity of fresh water abstracted at the existing water treatment plant at Plucks Gutter. In this way the water supply problems associated with direct re-use within a closed loop, i.e. effluent being re-used in the water supply to the same town (ref. 3) would also be avoided.

38. In order to establish the financial benefits from the use of the effluent, an existing mathematical model capable of calculating total cost for various resource options for the whole of Kent was re-run on three scenarios:-

(a) increasing the yield from Plucks Gutter by an amount equivalent to the discharge at Richborough of effluent from Ramsgate (15.9 ml/d)
(b) as (a) except with effluent from Ramsgate and Sandwich (town and industry) (25 ml/d)
(c) as (a) except with effluent from Margate, Birchington, Broadstairs and Ramsgate (46 ml/d).

39. The modelling took account of the resource costs to enlarge the treatment plant at Plucks Gutter and the necessary trunk mains to get the water into supply. The results obtained are set out in Table 2 and show that the maximum resource benefit is achieved with a flow of 25 ml/d, which would be provided by the effluent from Ramsgate and Sandwich, but the value is only £2.32M.

Table 2. Resource benefit

Effluent re-use from:	Flow:ml/d	Resource benefit: £ Million
Ramsgate	15.9	2.01
Ramsgate and Sandwich	25.0	2.32
Thanet	46.0	1.57

ECONOMIC APPRAISAL

40. The investigation considered a very wide range of options and each in sufficient depth to try and ensure that realistic estimates were available for comparison. The estimates, at mid 1982 price levels, included both capital and running costs, the latter being important in view of the continuing need to reduce revenue expenditure.

41. A number of options were resolved early in the investigation. For example there was a long standing proposal for combining Birchington with Margate by a scheme that provided sewerage improvements as well as abandonment of the Birchington outfall; this proposal was the cheapest option and Birchington was therefore considered to be part of Margate in the study. In addition there was no cost advantage in combining Minster with other areas and Minster was therefore excluded from further detailed comparison.

Capital Costs

42. Costs for STW's, outfalls and tunnelling were obtained from manufacturer's/contractors and checked against similar schemes adjusted, as necessary, for inflation. The cost models in TR61 (ref. 4) were found to be valid when checked against existing sewer and pumping station contracts and therefore used to estimate costs for sewers, rising mains, pumping stations and similar structures. Table 3 provides a summary of capital costs.

Table 3. Capital costs, mid 1982 prices

Outfall or STW	Capital cost of options: £ million						
	1	2	3	4	5	6	7
Foreness outfall	M 4.57	-	-	M 4.57	M 4.57	-	-
North Foreland outfall	B 4.59	M,B 11.57	M,B,R 22.60	B 4.59	B 4.59	-	M,B,R STW 24.75
Ramsgate outfall	R 11.27	R 11.27	-	-	-	-	-
Richborough STW	-	-	-	R 9.98	R,S 12.42	M,B,R 21.51	-
Sandwich STW	S 3.16	S 3.16	S 3.16	S 3.16	-	S 3.16	S 3.16
Total capital: £ million	23.59	26.00	25.76	22.30	21.58	24.67	27.91

M=Margate   B=Broadstairs   R=Ramsgate   S=Sandwich

Operating Costs

43. Total operating costs were evaluated to include power, employees, premises, supplies, services, transport and mobile plant.

44. Power costs were significant at some 50% of total costs and therefore evaluated separately for each part of each scheme to provide an accurate total.

45. A basic labour force, based upon mobile gangs, was evaluated for each option and costed in accordance with current Authority pay levels and overheads. A high degree of automatic control and "day shift only" working was assumed.

46. Costs for premises, supplies, services, transport and mobile plant were derived from typical Authority budgets. Sludge disposal costs were evaluated for each treatment works option and it was assumed there was no income and no charge for disposal of sludge to agricultural land. Table 4 shows the summary of operating costs.

Net Present Value Analysis

47. Financial analysis using the NPV method was carried out to give proper consideration to both capital and operating costs. Table 5 shows the summary of the NPV analysis.

48. Costs were discounted over the life of the asset with the longest life and asset lives were assumed to be 20 years for mechanical/electrical plant, 60 years for structures and 100 years for pipelines. The discount rate used was the Government's test rate of 5%.

49. It was assumed that there would be no significant change over the life of the schemes regarding labour costs and construction, plant and machinery costs.

Table 4. Operating costs, mid 1982 prices

Outfall or STW	Operating cost of options: £ million per annum						
	1	2	3	4	5	6	7
Foreness outfall	M 0.057	–	–	M 0.057	M 0.057	–	–
North Foreland outfall	B 0.059	M,B 0.136	M,B,R 0.285	B 0.059	B 0.059	–	M,B,R STW 0.425
Ramsgate outfall	R 0.067	R 0.067	–	–	–	–	–
Richborough STW	–	–	–	R 0.236	R,S 0.338	M,B,R 0.671	–
Sandwich STW	S 0.090	S 0.090	S 0.090	S 0.090	–	S 0.090	S 0.090
Total operating cost: £ million per annum	0.273	0.293	0.375	0.442	0.454	0.761	0.515

M=Margate   B=Broadstairs   R=Ramsgate   S=Sandwich

Table 5. Net present values, mid 1982 prices;5% discount rate

Outfall or STW	Net present value (NPV) of options: £ million						
	1	2	3	4	5	6	7
Foreness outfall	M 5.50	-	-	M 5.50	M 5.50	-	-
North Foreland outfall	B 5.63	M,B 13.74	M,B,R 27.78	B 5.63	B 5.63	-	M,B,R STW 32.87
Ramsgate outfall	R 11.84	R 11.84	-	-	-	-	-
Richborough STW	-	-	-	R 15.19	R,S 20.01	M,B,R 36.77	-
Sandwich STW	S 5.20	S 5.20	S 5.20	S 5.20	-	S 5.20	S 5.20
Total NPV: £ million	28.17	30.78	32.98	31.52	31.14	41.97	38.07

M=Margate   B=Broadstairs   R=Ramsgate   S=Sandwich

Area Scheme vs Separate Schemes

50. Area schemes for Thanet (Margate, Broadstairs and Ramsgate) were more costly than separate town schemes and it was found that transfer costs were a significant item in total costs. The following two examples illustrate the point.

51. First, compare the area long sea outfall for Thanet with the three separate town outfalls. The area scheme at capital cost £22.60 M was some £2.17 M more costly than the separate schemes and includes a transfer system cost of £4.9 M (21% of total) which outweighed any economies associated with building one outfall instead of three. The total operating cost of the area scheme at £0.285 M pa was some 56% higher and the power cost was found to be more than double the separate scheme cost. These combined to give a area scheme NPV some 21% higher at £27.78 M.

52. Secondly, compare the area STW for Thanet with the separate outfalls for Margate, Broadstairs and Ramsgate. The area scheme at capital cost £21.51 M was only £1.08 M more costly but the transfer system cost was £6.7 M (31% of total). The total operating cost at £0.671 M pa was some 367% higher, due mainly to treatment costs, but some 37% of total cost at £0.249 M pa was associated with transfer operating costs.

Marine Treatment vs Inland Treatment

53. Inland treatment was found to be more costly than marine treatment and the following two examples illustrate the point.

54. First, compare the area sea outfall scheme for Thanet at N Foreland with the area STW, at Richborough. The capital costs are similar at £21.51 M for the STW and £22.60 M for the outfall. The operating costs of the STW was, however, at £0.671 M pa, some 235% higher than the outfall scheme and contributed to the NPV of the STW being 32% higher at £36.77M. This demonstrates the significance operating costs may have on NPV calculations.

55. Secondly, compare the area sea outfall scheme for Thanet at N Foreland with the area STW also at N Foreland. The total capital cost of STW at £24.75 M is some £2.15 M more costly than the outfall and reflects the problem of attempting to balance treatment costs and outfall length. Outfall construction costs are not proportional to length as large fixed costs are associated with mobilisation and demobil-isation of marine plant/equipment. On recent contracts these fixed costs have been some 30% of the total contract price for outfalls in the 2 to 2.5 km range and the outfall length could have been increased by some 25% for a 10% increase in cost. Outfall costs are thus not significantly effected by small changes in outfall length.

56. The operating cost of the STW at £0.425 M pa was some 49% higher than the outfall but was less than expected. This was because the major cost was associated with transfer, which was common to both, and flows to the STW gravitated through the site and the outfall discharge. Extra power costs were principally sludge handling. The total NPV of the STW scheme at £32.87 M was some £5.1 M (18%) higher than the outfall solution.

Discount Rates

57. To check the effects of different discount rates, which may be significant (ref. 5), the options were appraised at rates of $2\frac{1}{2}$% and $7\frac{1}{2}$%. The relative order of options was unchanged but Table 6 illustrates the impact of different discount rates on options where operating costs are significant.

INVESTIGATION SUMMARY

58. The actual capital cost of the various options were very similar but the cheapest in terms of NPV was option 1 (Table

Table 6. Effect of different discount rates for Ramsgate options

Ramsgate option	NPV:£ million		
	$2\frac{1}{2}$%	5%	$7\frac{1}{2}$%
STW	25.49	15.19	11.76
Outfall	16.35	11.84	10.21

5) providing separate solutions for each town, separate outfalls for the Thanet coastal towns and a STW for the inland town of Sandwich. This reflects the impact of transfer costs mitigating against area solutions and the conclusion that marine treatment is generally the cheaper solution for coastal towns.

59. The chosen strategy was separate outfalls at Margate and Broadstairs and a STW at Ramsgate. Sandwich would receive works treatment and there would be further consideration of a combined works at Richborough serving Ramsgate and Sandwich.

60. Improvements at Ramsgate would be a costly part of any scheme and, although the works was marginally more costly, it was considered it would assist public acceptance of the whole scheme. The resource benefit of effluent re-use would halve the difference in NPV's.

61. The promotion of separate schemes would result in early improvement in the bathing water conditions as the completion of each scheme would result in an immediate reduction in pollution and bacteriological conditions on the coastline.

62. The area treatment options, both outfall and works, were more costly and would have the greatest impact on the Authority's capital programme due to the need to complete the treatment facility and transfer system before any benefit would be available.

PUBLIC CONSULTATION

63. The Margate Outfall was the first in England to follow the full procedure required under Part II of the Control of Pollution Act 1974. The main change was the requirement for a Discharge Consent from the Department of the Environment, the application for which had to be advertised in the local and national press.

64. If a long consent procedure was to be avoided, it was important to prevent objections to the advertised proposals, and a period of about three months was set aside to sell the proposed Thanet outfalls to the public. The form of these discussions and meetings has been fully described (ref. 6) but our experience has convinced us of the importance of this consultation stage in the progress of an outfall scheme.

65. The importance of local Members of Parliament and County and District Council Members in forming local opinion must be accepted and time must be spent in explaining the proposals and alternatives, if any. In the case of Thanet, the final public meeting was chaired by a senior local inhabitant and his neutrality contributed to the effectiveness of the meeting which accepted the proposals, with positive support from the two Members of Parliament and the Local Council.

66. This section moves away from the main topic of the paper, but the Authors believe their experience shows the value of public consultation and can help others over a potentially difficult hurdle. In the case of Thanet, the Margate Outfall was advertised in 1984 and that for Broadstairs in 1986. There were a total of three objections,

none of which were of a serious nature and Consents were granted by the Department of Environment based on the technical facts submitted in the applications.

Appendix 1

DESIGN ASSUMPTIONS AND STANDARDS
Population and Flow
1. Estimates of population, dry weather flow and sewage strengths together with forecasts of growth to 2001 were prepared. Works were designed for flows and loadings to 2001 and, owing to the small growth predicted, multi-stage construction was not considered.

Flow to Treatment
2. Design flows for full treatment at treatment works and for discharge to long sea outfalls were taken as peak dry weather flows i.e. 3 x (domestic + trade) + (infiltration)

Storm Overflows
3. Authority policy in respect of storm overflows is to pass flows up to approximately 6 DWF through the treatment works, where flows between approximately 3 and 6 DWF receive primary treatment prior to discharge to the receiving watercourse. This standard is based upon works discharging to watercourses and it is clearly inappropriate to pump storm flows from a sea outfall location, discharging to a large body of water, to a treatment works discharging to a river, which is a relatively much smaller body of water. It was also considered that this operation would tend to penalise the inland treatment options due to the extra transfer costs.
4. It was therefore assumed that the existing sea outfall locations would be retained and on-site storage of storm flows provided (ref. 7). This storage would retain the "first flush" of polluted storm water that is a feature of combined sewerage systems, for subsequent full treatment and also greatly reduce the frequency of operation of the storm outfall.

Treatment Standards
5. The Water Authority standards for long sea outfalls were adopted as follows: the discharge should be free from identifiable sewage solids, not cause a conspicuous slick, not cause objectionable conditions on the foreshore, not adversely affect flora and fauna and should satisfy EC standards for bathing waters. In addition the dissolved oxygen content and the concentration of toxicants in the receiving water more than 100 metres from the outfall should, respectively, exceed 4 Mg/l and be less than one tenth of the 48 hour LC50.

Outfall Headworks
6. All discharges to long sea outfalls to receive preliminary treatment of grit removal and fine screening, with

macerated screenings returned to flow. Headworks to be enclosed to minimise noise, smell and visual intrusion.

Sludge Disposal

7. Sludge to be disposed to farmland following pressing and storage.

REFERENCES

1. COUNCIL OF THE EUROPEAN COMMUNITIES. Directive of 8th December, 1975 concerning the quality of bathing water. Official Journal of the European Communities, vol.19, no.L31, 5th February, 1976.
2. IRVING T.E. and SOLBE J.F. Chlorination of sewage and effects of marine disposal of chlorinated sewage: a review of the literature. Water Research Centre, March 1980, Technical Report TR130.
3. ROBERTS G.D.M. and COWAN J.P. Sewage effluent as a resource. Symposium Proceedings on Water Resources - change in strategy, Institution of Civil Engineers, London, 1979.
4. COST INFORMATION FOR WATER SUPPLY AND SEWAGE DISPOSAL. Water Research Centre, Technical Report, TR61, November 1977.
5. CLOUGH G.F.G. and CANNON D.E. Principles of economic comparisons. Conference Proceedings on Coastal Discharges - engineering aspects and experience, Institution of Civil Engineers, London. 7-9th October, 1980.
6. MIDMER F.N. and WELSH P. Gaining public support. Water Bulletin, 10th May, 1985, no. 157,8-9.
7. Technical Committee on Storm Overflows and Disposal of Storm Sewage. Final Report. HMSO, 1970.

12. An economic assessment of inland treatment and marine treatment options for Weymouth and Portland

R. TYLER, BSc, MICE, MIWES, Divisional Agency and Special Projects Engineer, Wessex Water Authority

SYNOPSIS. Wessex Water Authority has constructed a long sea outfall to provide sewage disposal by marine treatment for Weymouth and Portland. Before construction commenced an economic evaluation of marine treatment and alternative inland treatment options was carried out. Post construction experience enables previous assumptions to be examined critically.

INTRODUCTION

1. Shortly after the formation of Wessex Water Authority in April 1974, a critical examination of the worst problems inherited showed that the problems of sewage disposal in Weymouth and Portland featured high in the list of priorities. In Weymouth, a sewage pumping station built in the 1890's discharged sewage through short sea outfalls only 200 m from the Weymouth bathing beach area. The old twin pumping mains passing through the town centre and harbour areas were subject to frequent failure, resulting in considerable pollutions of the harbour and Backwater (a marina area). Pumping capacity was less than 3 x current DWF, flows in excess being discharged to the Backwater. In Portland, sewage was disposed through a multitude of short sea outfalls, some barely extending to low water level.

2. The Authority undertook to give high priority to solving the problems in conjunction with the newly formed Weymouth and Portland Borough Council, and appointed Consultants, John Taylor and Sons, to advise on sewage treatment matters. L G Mouchel and Partners continued their previous association with the town and were appointed to assist the Council develop the necessary new sewerage system.

3. Early desk studies indicated that a new long sea outfall discharging to the west of Portland in West Bay was likely to be the best technical (and economic) solution, since the water in West Bay was deeper and had better tidal characteristics than Weymouth Bay, and there were few obvious sites for inland treatment works. The Consultants were instructed to proceed with detailed studies to determine the most suitable discharge point for an outfall, culminating in a Final Report in September 1975, recommending an outfall

which could serve both Weymouth and Portland discharging in West Bay, approximately 1 km offshore.

4.   The Final Report, based on extensive offshore investigations, effectively confirmed the ideal location for the outfall dispersion point, but the ancillary works necessary to re-drain Weymouth and Portland were only considered in outline at this stage. A sectionof the Report dealing with alternative inland treatment options indicated overwhelming technical advantages for the marine treatment option and concluded that there were likely to be significant cost advantages also. Subsequently, the Consultants amplified their report on inland treatment options, and following consideration by the Authority the recommendation for marine treatment was approved. Accordingly detailed design of the outfall, headworks and associated sewerage works was allowed to proceed.

5.   During the next three years the detailed design of the project was developed, and for various reasons it underwent many changes from the outline proposals envisaged in 1975/76. The route of the outfall tunnel itself changed significantly, as did many of the proposed trunk sewers and rising mains in Weymouth and Portland, whilst the headworks changed from a simple comminutor installation to a sophisticated pretreatment plant in an environmentally acceptable enclosed building. Because of the significant changes that had taken place since the scheme's conception, the Authority called for a report to reassess the merits of the proposals against inland treatment before finally committing itself to the marine treatment solution. This report was commenced during the tendering period for the outfall and headworks, and concluded shortly afterwards, with the benefit of the tender prices. This report in January 1979 concluded that the combined marine treatment scheme for Weymouth and Portland was economically superior, and accordingly the Authority authorised work to commence. Construction of the tunnelled outfall was finally completed in July 1983, and commenced treating sewage in January 1984 when the flow from Weymouth was successfully diverted. Flow from the Wyke Regis area was added in April 1985 and flow from the Underhill area of Portland is to be included early in 1987, with the commissioning of the Victoria Square Pumping Station. Further contracts will extend the catchment to include the remainder of Portland, and the scheme will be complete by 1990. The performance of the outfall has been extensively monitored subsequent to commissioning and has been found entirely satisfactory.

THE 1976 ASSESSMENT

6.   Consultants, John Taylor and Sons, examined 7 sites, (5 in Weymouth and 2 in Portland) as possible inland treatment works locations, against a 13 point list of features with which the ideal site would conform.

Receiving Waters

7. It quickly became clear that by far the most important of these was the need for an inland treatment site to have in the close proximity receiving waters of sufficient quantity to be able to accept the treated sewage works effluent, and larger quantities of partially treated storm sewage. The largest watercourse in Weymouth is the River Wey, but the base flow, (in summer approximately 0.15 cumec), is very small for a possible recipient of treated effluent. Mass balance calculations quickly established that treated effluent would have to be consistently less than 5 mg/l BOD Ammonia and Suspended Solids, a standard which is impractically high, and furthermore that the River Wey was incapable of coping with storm flows without causing flooding.

8. For different reasons, the Fleet was also considered unsuitable. The Fleet is a shallow tidal water between the mainland and Chesil Beach, with a single connection with the sea at Portland Harbour. Continuous discharge would not be possible, and storage of 8 hours peak flow would be necessary at the treatment works. For all inland treatment options, therefore, it was concluded that treated effluent and storm sewage would have to be discharged to sea. Desk studies indicated that an effluent outfall in West Bay would need to extend approximately 400 m, but a long outfall approximately 1.2 km would be needed in Weymouth Bay to obtain a sufficiently high initial dilution.

Treatment Process

9. Only the activated sludge process was considered suitable because the sites were closer than desirable to development. Outline costings were based on conventionally designed activated sludge plants with surface aeration, designed to future 2010 A.D. population loadings, with sludge treatment by heated digestion and disposal by tanker to farmland in line with Wessex Water Authority practice.

Preliminary Economic and Financial Conclusions

10. One site in Weymouth and one in Portland did not merit detailed economic analysis due to their technical shortcomings after a more detailed investigation. The remainder were considered as possible sites for combined treatment works or for separate works for Weymouth and Portland. It was concluded that a combined inland treatment option for Weymouth and Portland would be cheaper than separate works for each area, the best works site being near Martleaves Farm. The sewerage works necessary for this option would be broadly similar in cost to the marine treatment scheme, but the capital costs of the inland treatment works and shorter effluent outfall were likely to exceed the marine treatment costs by more than 100%. Overall, the capital cost advantage of the marine treatment option was likely to be between 30 and 50%. Operating costs for inland treatment would clearly exceed those for marine treatment by a significant margin.

Table   1.   1979   Comparison   of   Options

	Capital Costs £m Q3/78					
Option	STW or Head-works	Efflu-ent Pipe & Outfall	Sewer-age Works	Total Capital	Range of Uncer-tainty	Total Fore-seeable Cost
1. Weymouth & Portland combined STW at Martleaves Farm				22.83	2.24	25.07
1. Ditto at Wyke Castle				22.97	2.25	25.22
3. Weymouth STW at East Fleet & Portland at Sweet Hill				18.05	2.95	21.00
				7.14	1.06	8.20
				25.19	4.01	29.20
4. Weymouth STW at Mount Pleasant Farm, Portland at Sweet Hill				20.63	3.79	24.42
				7.14	1.06	8.20
5. Marine Treatment for Weymouth & Portland with Head-works at Wyke Regis	1.49	3.38	8.34 4.38 12.72	17.59	0.61	18.20
6. Marine Treatment for Wey-mouth only with Head-works at Lodmoor, inland STW	2.34	7.40	11.27	21.01	3.62	24.63

Table 1.   1979 Comparison of Options (Continued)

Option	Capital Costs £m Q3/78					
	STW or Head-works	Efflu-ent Pipe & Outfall	Sewer-age Works	Total Capital	Range of Uncer-tainty	Total Fore-seeable Cost
for Portland at Sweet Hill				7.14	1.06	8.20
				28.15	4.68	32.83
7. Marine Treatment for Weymouth and Portland with Headworks at Furzedown Farm	1.49	5.26	9.34 5.93	16.09 5.93		
				22.02	2.11	24.13
8. Marine Treatment for Weymouth and Portland with Headworks at Wyke Bridging Camp	1.49	4.85	9.44 5.68	15.78 5.68	1.52 0.54	
				21.46	2.06	23.52

11.   Since both capital and operating costs of the marine treatment option were estimated to be significantly lower than the best inland treatment option, further economic or financial analysis was not carried out at that time, as it was concluded that the case for marine treatment had been conclusively demonstrated. The usual financial disadvantage to marine treatment options is the need to invest large capital sums in the early years to enable the asset to be utilised.   However, in this case, since both marine and inland treatment options involved major new construction sites with trunk sewers and pumping stations to redirect sewage from the existing outfalls, considerable capital investment was required in the short term whichever option was selected, and there was no financing cost advantage to the inland treatment option as may be the case when phased extensions of an existing treatment works is being compared with a new sea outfall.

THE 1979 ANALYSIS
12.   The 1979 report considered the costs of the long sea outfall scheme as designed and ready to be constructed, together with alternative marine treatment options and a review of the inland sewage treatment works options previously considered in 1976. Because the combined Weymouth and Portland outfall scheme had been developed to a greater degree of detail than the other options, it was felt that costings for this option should be more reliable, and this should be reflected in comparison with other options. Accordingly, wherever possible an upper limit of possible expenditure was quoted to indicate any difference in reliability.   The summary of likely capital costs as estimated in January 1979 is shown in Table 1 below. Operating costs were not evaluated in the 1979 Analysis.
13.   The 1979 reassessment again confirmed that the design scheme (Option 5) was considerably superior in capital cost terms, being £5.32m (29%) cheaper than the next best option, (an alternative marine treatment scheme), and £6.87m (38%) cheaper than the best inland treatment option.

THE PRESENT POSITION (EARLY 1987)
14.   With the completion of the headworks and outfall in 1983, and the connection of Weymouth early in 1984, approximately 65% of the scheme had been completed.   Since then the Wyke Regis area has been added by means of a temporary sewer connection.   The new Wyke Regis Pumping Station is approximately 50% complete, and when the new pumping station at Ferrybridge is also constructed all main drainage works in Weymouth will be complete.   The pumping main from Portland to the headworks has been laid, and the new Victoria Square Pumping Station at Underhill is ready for

commissioning.   Trunk sewers in the Underhill area of
Portland have been laid, but the main gravity sewer around
the eastern coast of the island to connect the remaining
population will not be complete until 1990.
15.   Expenditures already incurred and projected to
completion are shown in Table 2, both at out-turn and at the
price base adopted for the project, third quarter 1978
prices, (the price base for the outfall and headworks
tender).   The 1979 Estimates are also shown for comparison
with the latest financial position.

Table 2. Expenditure Forecasts to Completion

	Capital Costs £m			
	Actual Costs/ Latest Predictions		1979 Estimates	
Works	Outturn	Q3/78	Estimated Cost Q3/78	% Increase Latest Over 1979
Headworks	3.000	2.105	1.49	41
Outfall & Diffusers	8.404	5.840	3.38	73
Total Marine Treatment	11.404	7.945	4.87	63
Weymouth Sewerage (a) Completed (b) Outstanding	11.547 0.764	8.504 0.390		
Total	12.311	8.894	8.34	7
Portland Sewerage (a) Completed (b) Outstanding	5.698 5.892	3.216 2.746		
Total	11.590	5.962	4.38	36
Total Sewerage	23.901	14.864	12.72	17
Total Marine Treatment & Sewerage	35.305	22.810	17.59	30

16.   With the project more than 80% complete in financial
terms, the latest cost predictions are considered to be
reasonably accurate, and it is interesting to draw
conclusions from the comparisons with the earlier estimate.
Since the scheme reached substantial detail design stage
(1979), further cost increases of 63% (marine treatment) and
17% (sewerage) have been or are likely to be incurred in
bringing the project to completion.

17.   It is noteworthy that by 1979, the sewerage design
works for the Weymouth connection were substantially
complete, whereas the Portland design was only just being
developed from the preliminary proposals. This is clearly
the reason for small cost increases in the Weymouth sewerage
proposals (7%), whereas a significant cost increase in the
Portland sewerage proposals (36%) shows the difficulty of
estimating major schemes accurately at the preliminary stage
when many further investigations need to be done. The 63%
increase in the headworks and outfall costs is mainly due to
serious constructional problems encountered during the
undersea tunnelling work. Major changes to the design of
this tunnel and diffuser system had to be developed to
overcome these problems, which have been well documented in
previous papers, (refs 1-4). The apparent difficulty in
producing reliable cost estimates at the early stages of a
large project such as this is considered to be due to two
main factors
(a)   the uniqueness of the project.
(b)   the significant element of tunnelling work involved,
      particularly the undersea work.

18.   Whereas cost estimates for civil engineering works of a
more routine nature can usually be prepared with some
confidence based on recent previous project costs, there are
seldom many similar schemes from which to draw experience for
the larger, "one-off" projects. What information is
available may not be recent tendering experience, or in
directly comparable conditions. The uniqueness of the
project also prevents the engineer from fully comprehending
all the various factors which will become apparent during the
detail design, usually requiring significant changes from the
preliminary proposals, and which inevitably cause cost
escalation rather than cost reduction. Similarly, experience
with tunnelling works generally indicates that considerable
cost increases may often be expected, due to the lack of
flexibility in working if problems are encountered. This was
particularly evident on the undersea tunnel, whose entire
2.7km length had to be constructed on a single onshore access
shaft.

19.   The Wessex experience on the Weymouth and Portland
project indicates that appropriate risk factors should be
considered for large projects if a proper appreciation of the
likely capital costs is required. Preliminary estimates are
likely to be subject to considerable variation as the project
is developed, and by the time that the first major contract

is let. The risk factor to be used in schemes may be varied depending on the proportion of sewerage in the scheme, (with its lower risk of cost escalation), depth and size of sewers, and the presence or absence of tunnelling works in the proposals.

OPERATING COST COMPARISONS

20.   It was always known that the operating costs of the sewerage system would be very similar, whichever treatment option were chosen, and these were not therefore assessed in detail in 1979. Significant savings were expected however for the operation of the headworks over a full inland sewage treatment works. As the design of the headworks developed, it changed from a simple comminutor installation to include grit removal and fine screening to 5mm aperture size, with all the associated maceration and washwater straining machinery requirements. Because of the proximity of housing development to the headworks, forced ventilation with activated carbon odour destruction was incorporated. The design requirement was for a fully automatic fail safe headworks which would generally operate unattended, and in 1979 it was expected that running costs would be between £40,000 and £45,000 per annum when operating at design flow rates.   These costs are approximately £58,000 to £65,000 at 1985/86 outturn values.

21.   The headworks however, have now been operational for nearly three years, and it is possible for actual running costs to be evaluated. Full year figures for 1985/86 may be summarised as follows

Table 3. Headworks Operating Costs 1985/86

		£000's
(a)	Energy (electricity and fuel for standby generator)	19.8
(b)	Odour destruction (activated carbon)	4
(c)	Drum screen spares and chemical degreasants	5.5
(d)	Macerator spares	12
(e)	Instrumentation and minor equipment maintenance	2.2
(f)	Ground and building maintenance	4
(g)	Labour, transport and miscellaneous	33
	Total	80.5

22.   Points of note from the above are that unexpectedly high expenditures have been incurred on replacement parts for the drum screens and macerators. Nine new panels had to be purchased for the drum screens at approximately £500 per panel when the 8 mm thick plastic screening medium developed cracks. Design modifications by the screen manufacturer, and more stringent fixing procedures now introduced are expected to overcome these problems. Similarly, measures are being taken which should reduce the high maintenance costs of the macerators, which have to break down screenings to a nominal 3 mm size, (4 mm when the macerator wears), for solids to pass through the   5 mm diameter apertures in the drum screens. To achieve this degree of fineness, double maceration is currently employed with 4 macerators (2 duty, 2 standby). It has been found in practice that replacement cutting parts for each pair of macerators are required after four months operation, at a cost of £2,000 per macerator. Although the manufacturer attempted to extend the life of the cutting parts by improved hardening and manufacturing techniques, no significant reduction in the cost of the spares has been possible until this year, with the development of a new macerator by the original macerator supplier. One of these macerators is currently on trial at Weymouth, and although it is too early to predict the operational life of its cutting blades, the spare parts only cost about £300 per macerator, and strip down time for the new macerator to replace the blades is significantly less. Thus it is expected that future macerator maintenance costs are likely to be much less than the 1985/86 values. The power consumption of the new macerators is approximately half that of the existing, and accordingly it is considered that by 1987, power and labour costs should be reduced from the 1985 values by virtue of the new macerating equipment. Ground maintenance costs were unusually high during 1985/86 when it was necessary to replace a significant number of the shrubs planted around the site.   Once established, site maintenance should reduce substantially.

23.   Current flow through the headworks is approximately 13.4 Ml/d DWF, (averaged over the year) with an average daily flow of 16.8 Ml/d allowing for storm discharges, compared to an estimated 19.3 Ml/d DWF for the entire Weymouth and Portland conurbations by 2010 AD in the 1976 Report. By 2010 it is reasonable to expect all flow related operational costs (indicated * in Table 4 below) to be increased pro rata, i.e. 44%.   Thus, after making appropriate adjustments to items (a), (c), (d), (f) and (g), to take into account economies anticipated by the measures described in paragraph 22, it is possible to estimate future operational costs as follows

Table 4.   Present and Future Estimated Headworks
Operational Costs

		£000's 1985/86 average out-turn		
		1985/86	1985/86 Adjusted	2010
(a)	Energy *	19.8	18	29.5
(b)	Odour destruction *	4	4	5.8
(c)	Drum Screen *	5.5	1.5	2.2
(d)	Macerator *	12	1.5	2.2
(e)	Instrumentation, etc.	2.2	2.2	2.2
(f)	Ground & building maintenance	4	1.5	1.5
(g)	Labour, transport	33	32	32
		80.5	60.7	71.8

24.   It can be seen that the best comparison available with
design stage estimates of operating costs show that actual
costs are some 10-20% higher than expected.   Considering the
difficulty of accurately estimating the degree of attendance
necessary to fully service the plant at the design stage,
the accuracy of the estimate is considered reasonable.
25.   Despite the escalations in operating costs for the
marine treatment headworks, they are still very much less
than the likely operating costs of an equivalent inland
sewage treatment works.   Although such costs vary according
to the treatment process used and the method of disposing
of sludge, a typical operating cost for a Wessex Water
Authority treatment works of equivalent size is about
£190,000 per annum, i.e. approximately 200 % greater.

IMPACT ON CAPITAL PROGRAMME
26.   Any major scheme of marine treatment is likely to
involve considerable capital investment, and there is
immediately a dilemma in that any expenditure before the
outfall becomes operational can be regarded as wasted
economic resources, whereas a concentrated construction phase
to minimise this waste of resources results in a high demand
for capital allocations immediately prior to commissioning.
It may be necessary to phase the works in advance to produce
a projected cash flow which does not have an unacceptable
impact on the Water Authority's capital programme, although
the options for phasing on the marine treatment works are
generally limited.

27.   Although from an economic viewpoint phased works are
disadvantageous, there are often practical advantages.   A
rolling programme of design and construction work can be
prepared which means that the same design team can move from
one phase of the project to the next with the benefits of
experience gained from the previous phase.   Similarly,
construction works can be broken down into reasonably sized
contracts which enable local contractors, who are usually
more competitive, to tender without being over committed, as
may be the case if, say, all the sewerage works are let as
one large scheme.   Indeed, the likely value of the work if it
is let as a single contract may preclude local contractors
from tendering, and higher costs overall may result.   In the
final analysis, it is a matter of judgement for each
Authority to make, but in the case of the Weymouth and
Portland project, it was decided at a fairly early stage that
the sewerage works would be phased into contracts of
approximately £1m to £2m in value, whereas the complete
marine treatment scheme, (outfall, diffusers, headworks and
machinery) would be let as a single contract.   The contract
period necessary to construct the sea outfall tunnel
effectively provided a phased investment.

CONCLUSIONS
28.   From experience gained on the Weymouth and Portland
Marine Treatment Scheme, it is possible to draw the following
conclusions

(a)   A properly designed and executed scheme for the
        disposal of sewage by marine treatment provides an
        alternative solution to inland treatment methods which
        can be economically and environmentally superior.   This
        has been demonstrated by a regular monitoring programme
        of both beaches and the marine environment adjacent to
        the diffusers, and the complete absence of smell or
        noise problems from the headworks.
(b)   An economic analysis of marine treatment versus inland
        sewage treatment options should be carried out as part
        of the overall decision making which will take into
        account technical, environmental and political issues.
        Discounted cash flow analysis will indicate the
        preferred economic solution, but will only be useful if
        the capital costs are lower for the inland treatment
        option, since operating costs for inland treatment are
        invariably higher.   The analysis should include
        sensitivity testing to likely increases in capital and
        operating costs.
(c)   A financial analysis should be carried out if there
        appears to be a significant difference in the capital
        allocations required in the short, medium and longer
        term.   The preferred option from the financing cost
        aspect may differ from the most economic option
        indicated by lowest net present value, and the
        financial impact of both options should be considered

in relation to the Water Authority's capital works programme.
(d) Requirements to achieve environmental and technical objectives may change during the timescale needed to develop a major scheme, and final solutions may be significantly different in concept and cost from proposals considered at the conceptual stage. Risk factors appropriate to the nature of the scheme should be used to enhance preliminary estimates to provide more accurate predictions of likely final capital and operating costs.
(e) Marine treatment works are more likely to increase in cost than associated sewerage works, particularly for a tunnel outfall and appropriate risk factors should be used to reflect this.

REFERENCES
1. KEMBLE J. R. and YOUNG J. A. Weymouth and Portland marine treatment scheme - introduction history and options. Proceedings of the Institution of Civil Engineers, Part 1, 1984, vol. 76, February, 81-94.

2. PITTMAN J. F. H., MARTIN C. and KING M. W. Weymouth and Portland marine treatment scheme - inland sewerage works. Proceedings of the Institution of Civil Engineers, Part 1, 1984, vol. 76, February, 95-116.

3. ROBERTS D. G. M., FLINT G. R. and MOORE K. H. Weymouth and Portland marine treatment scheme - tunnel outfall and marine treatment works. Proceedings of the Institution of Civil Engineers, Part 1, 1984, vol. 76, February, 117-144.

4. ROBERTS D. G. M. and COOKMAN I. J. R. Coastal discharges - engineering aspects and experience. Pretreatment, screening and detritus removal, pp 81-88 Thomas Telford, London, 1981.

13. Sewage sludge to sea — the Thames Water approach

M. K. GREEN, BSc, Corporate Policy Manager, and
M. J. HANBURY, BSc, PhD, Senior Operational Research Analyst,
Thames Water Authority

SYNOPSIS

Sewage sludge production and disposal to sea from London commenced in 1889 and the practice has continued for almost 100 years. The present sludge disposal operation run by Thames Water takes 4.2 million wet tonnes of sludge from four works, Beckton, Crossness, Deephams and Riverside which together serve a population of 4.4 million people. Four ships take the sludge to the Barrow Deep at an annual cost of £4.5 m.

Thames Water considers the first step in responsible sludge disposal is the control of pollutants from entering the sludge by effective trade effluent control which has shown dramatic results in recent years. Secondly Thames fully supports the concept of Best Practicable Environmental Option as defined in the Oslo Convention rather than the "Anticipation Principle" approach of the proposed EC Directive and indeed Thames consider that its present sea disposal operation is the Best Practicable Environmental Option for sewage sludge from London.

Thames Water's approach to sludge disposal is to continuously seek the most managerially responsible disposal route. A recent evaluation covering sea, land, incineration and reuse has identified some eighteen options which are thought to be technically feasible, practicable and operationally secure. From London, conventional land disposal is not practical leaving only sea disposal, landfill and incineration as viable options. Of those, the sea disposal options are consistently the most economic, land disposal costing approximately 1.5 times and incineration costing twice as much. Sea disposal can be substantially improved by thickening of the sludge thus minimising transport costs.

INTRODUCTION

1. The disposal of sewage sludge is both an emotive subject and an inescapable operation. For many years the pro's and con's of disposal to land have been debated with more and more stringent controls being applied. The ever increasing environmental awareness has seen the discussion extend to the practice of disposal to sea culminating in the recently published draft EC Directive.

2. Sludge has been disposed to the North Sea from London for nearly 100 years, the initial stimulus being the removal of pollution from the River Thames. In 1856 the Metropolitan Board of Works was established and under their jurisdiction interceptor sewers were constructred, completion being in in 1874. However even the discharge of untreated sewage below London created problems and in 1882 a Royal Commission was appointed to inquire into, and report upon, the system under which sewage is discharged into the Thames by the Metropolitan Board of Works. Their report in 1884 concluded that untreated sewage should not be discharged to the river. The solid matter should be removed and either disposed of to land or sea and the liquid portion discharged to river. Hence sludge was taken by steamer and dumped in the Black Deep. Disposal to sea commenced in 1889 from Beckton and 1891 from Crossness.

3. The issues involved in sludge disposal fit well into the concepts of River Basin Management, on which the Water Authorities were established in 1974, and Best Practicable Environmental Option, which is practised in the UK.

4. A recent House of Lords report[1] has supported the UK view on waste disposal to sea and has found against the need for an EC Directive.

5. This paper outlines the Thames position and some of the initiatives that are taking place in order to responsibly minimise the problem of sludge disposal.

PRESENT OPERATION

6. Figure 1 shows that the Thames Water area together with the catchment of the four works which now dispose of sludge to sea . The areas of the four catchments are small in relation to the whole of Thames, however they serve some 4.4 million people which is approximately one third of the total population served by Thames. The population served by these catchments is not expected to change in the foreseeable future.

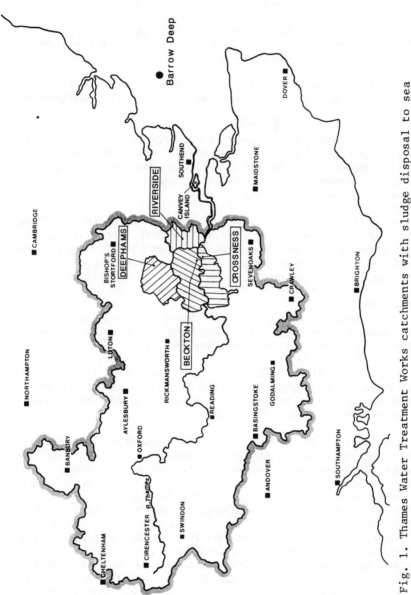

Fig. 1. Thames Water Treatment Works catchments with sludge disposal to sea

205

7. Four sludge vessels undertake a combined operation from Beckton and Crossness to the Barrow Deep disposal site which is some 15 miles south east of Clacton-on-Sea and some 55 miles from Beckton, in the outer Thames estuary.

8. The three older vessels, MV Newham, MV Bexley and MV Hounslow can carry 2300 tonnes. The newest of the fleet MV Thames can carry 2600 tonnes.

9. The ships sail on the ebb tide and return on the flood tide. There are therefore some 2820 possible sailings per annum.

10. The older ships have twelve officers and ratings and the newer one thirteen. Each ship has two crews which with the standby crew results in 126 officers and ratings. In addition there are six other managerial and admin staff associated with the marine group.

11. The total cost of sludge disposal to sea in 1984/5 was some £4.5 m and 4.2 million wet tonnes of sludge was disposed of with the fleet operating at 63% efficiency. The main reason for loss of sailings was crew shortage, although bad weather and maintenance had a significant effect.

CONTROLS

12. During most of the one hundred years that sludge disposal to sea from London has been carried out, there has been no legislation covering this activity.

13. The earliest control, which commenced in the early 1960's, was a voluntary scheme of authorisations of disposal. This scheme was operated by the Ministry of Agriculture, Fisheries and Food.

14. During the 1970's there were a number of International Conventions relating to the disposal of wastes at sea. The more important ones, when considering sludge dumping, are the Oslo, London, and Paris Conventions. Each of these conventions categorised pollutants into groups: List 1 or the black list contains pollutants which, because of their toxicity, persistance or bio-accumulation, are banned; List 2 or the grey list contains pollutants that exhibit similar but less marked characteristics than list 1 pollutants and these must be disposed of with special care. Included in this category are substances such as mercury, cadmium and organochlorine compounds. Included in List 2 are substances such as copper, lead, cyanide and fluoride. Wastes which contain neither List 1 nor List 2 pollutants or contain only "traces" require a permit or licence for dumping. "Trace" was never however defined.

15. The criteria for allowing disposal is the concept of Best Practicable Environmental Option (BPEO) and in the UK MAFF execute its responsibility under the Conventions ensuring disposal is in accordance with the BPEO.

16. The underlying premise of the BPEO is that each waste has to be considered on its own merits so that the suitability of marine disposal can be judged against the alternatives for that particular waste.

17. In the UK in 1974, the Dumping at Sea Act was enacted to control disposal of wastes at sea. This was replaced in 1985 by the Food and Environmental Protection Act. MAFF was responsible for the enforcement licensing and monitoring under the 1974 Act and is responsible under the 1985 Act.

18. The removal of List 1 and List 2 pollutants from sludge is not economically viable and indeed in most cases is not technically possible. The only route therefore is to reduce these pollutants using existing trade effluent controls supported by the relevant legislation. The trade effluent control methods implemented by Thames Water have resulted in marked improvements between 1974 and 1985. The nickel, cadmium and mercury content of Beckton sludge is shown in Table 1. Reductions to a lesser extent with the other heavy metals have also been achieved. Improvement has also been made in the heavy metal content of Crossness sludge during this period.

Table 1. Metal Content of Beckton Sludge

	CONCENTRATION mg kg^{-1}	
	1973	1985
Nickel	300	150
Cadmium	61	27
Mercury	77	10

19. However, due to the large proportion of metals from diffuse sources, improvement of industrial effluents alone will not eliminate metals from sewage sludge. In areas such as London, there is little scope for further improvement of the sewage sludge by this route.

IMPACT

20. Attempts have been made at estimating the sources and polluting loads entering the North Sea. As the North Sea is bordered by several different countries each with their own interests to protect, this has not been an easy task. In addition the North Sea is an open sea therefore assessing the true impact of polluting loads is also difficult. Pollution, especially atmospheric, is neither a respector of national boundaries nor does it appear with a country of origin tag. Further problems are created by a genuine absence of data.

21. Bearing in mind all the above comments, the general conclusion drawn from the various studies[2,3], is that the major sources of pollution of the North Sea are due to rivers, dredging spoils and aerial deposition. It is estimated that pollution of the North Sea due to sewage sludge is minimal. If heavy metals are used as the indicator of pollution then the disposal of sewage sludge, industrial wastes and marine incineration account for only 5%[3]. This figure would be even lower if atmospheric deposition were also included.

22. In terms of the biological and nutrient load associated with sludge dumping it has been estimated that, excluding atmospheric deposition, sewage sludge accounts for 5% of BOD_5, 1% of nitrogen and 3% of phosphorus[3].

23. There is evidence that on the East side of the North Sea the incidence of disease in fish has increased and this appears to be associated with waste inputs. However the International Council for the Exploration of the Sea, a distinguished body on which all the fishery research laboratories of western Europe are represented, has concluded that without more positive evidence it is impossible to say whether dumping is responsible for fish disease.

24. The United Kingdom is now the only country disposing of sewage sludge to the North Sea and Thames Water's sea disposal operation accounts for 80% of all sewage sludge in the North Sea.

25. Studies of the impact of the sludge disposal activities of Thames Water and its antecedants date back to the 1890's and there have been 26 major studies since 1955[4].

26. Two reports by MAFF[5],[6] in the early 1980's indicated that the sludge dumping activities of Thames Water could be detrimental to the marine environment. During 1983/4 Thames carried out extensive surveys of the dump site and surrounding areas and was unable to substantiate the earlier claims of MAFF. This situation was obviously highly unsatisfactory.

27. Consequently in April 1985 Thames in conjunction with WRc commenced a four year detailed study of the environmental status of the Thames estuary and the effects of sludge dumping. Preliminary results of this study will be available in early 1987.

28. The stated objective of the study is to assess the environmental status of the Thames estuary and the effects of sludge inputs in particular. This will be achieved by:

(i)   Dispersion studies to quantify the distribution of dumped sludge solids;

(ii)   Studies of sediment quality with respect to the distribution of metals and organic contaminants;

(iii) Assessment of sublethal stress in mussels;

(iv)   Monitoring in indigenous species;

(v)   Bio-assay tests;

(vi)   A study of all inputs of contaminants to the tidal Thames;

(vii) A study on the effects of sludge disposal on fish in the Thames estuary.

THAMES WATER POLICY
29. Thames Water is of the firm belief that the most sensible and responsible method of sludge disposal from London is via the current disposal route ie to the North Sea. Thus Thames not only supports the concept of BPEO but is of the opinion that the current practice is indeed the BPEO.

30. This view is reached after a careful consideration of the logistics and environmental effects of the current practice and the other options available.

31. Thames Water's objectives concerning the disposal of sludge are:

(i)   to operate as effectively as possible whilst maintaining security of operation;

(ii) to operate as economically as possible thus ensuring good value for money;

(iii) to safeguard the environment.

32. It is management's task to pursue all three objectives which in practice means that there must be a balance between effectiveness, efficiency and environmental impact.

33. For the correct balance to be achieved, considerable information must be gained in order to make responsible managerial decisions on the best course of action. To this end, research in Thames is geared not to defending the current practice but at investigating without prejudice the effects of the current operation and the feasibility of alternatives. This enables continuous assessment of which method of disposal is the BPEO.

34. Thames' research and that of others has shown that the effect of sewage sludge on fauna in the Thames Estuary is very slight and can only be detected with great difficulty. This is not unexpected as the North Sea has a considerable capacity to receive wastes without causing pollution. Also for some categories of wastes, when compared to the other land based options, it is clear that the North Sea is in fact environmentally the best suited for disposal.

EC DIRECTIVE
35. Currently there is proposed new legislation on sludge dumping in the form of the EC Directive on the Dumping of Waste at Sea. The purpose of the Directive is to prevent and reduce marine pollution caused by dumping of waste at sea from ships and aircraft.

36. The Directive promotes the development of alternative land-based methods of disposal and encourages waste recycling before considering dumping at sea.

37. The Directive categorises pollutants in a similar way to the Oslo and London Conventions and seeks to reduce the amount of waste disposed to sea by: prohibiting the dumping of substances listed in Annex I (similar to List 1 in the conventions), reducing by half the dumping of substances listed in Annex II (similar to List 2), and requiring a general permit for the dumping of any other kind of waste.

38. For the first time "trace" contaminant has been defined as "contaminants occurring in quantities which have been shown by specific evidence to be unlikely to contaminate the marine environment". This definition is however not particularly helpful because the specific evidence is not available for the contaminants.

39. Many sections of the Directive are unclear as to their meaning and consequently parts may be interpreted in different ways. The wording of some parts will need to be resolved.

40. If, as the Directive may be interpreted, sewage sludge is included in Annex I then sludge dumping will be banned. If sewage sludge is exempt from Annex I by virtue of the clauses contained in Annex IB then there will be a requirement for a 50% reduction in the amount of sludge disposed of. This reduction was proposed to be effected in the period 1990-1995 at a rate of 10% per annum.

41. The adoption date of the proposed Directive is July 1986, with Annex I and II restrictions coming into force in 1990. The earliest adoption date is now thought to be towards the end of the decade with the restrictions commencing in the early to mid 1990's.

42. The recent House of Lords Select Committee[1] on the European Communities report on Dumping of Waste at Sea concluded that:

- pollution is mainly caused by aerial deposition and by rivers with sewage sludge contributing only a minor amount;

- the total input of heavy metals to the sea from dredging spoils is at least ten times that from sewage sludge;

- there is scant scientific evidence of pollution attributable to present dumping practices;

- sea disposal is the Best Practicable Environmental Option;

- it is inappropriate to draw up a single regime for dumping in seas as different as the North Sea, the Atlantic, the Baltic and the Mediterranean;

- the proposed Directive is unnecessary and ill-conceived.

THE ALTERNATIVES

43. It has been shown that the future of sludge disposal from east London is unclear and this situation has prevailed for some time. With the threat of sludge dumping being curtailed Thames Water decided in 1984 to evaluate the options available as alternatives to and for improvement of the existing operation.

44. This study considered various thickening processes for both raw sludge and digested sludge in conjunction with sea disposal and various land disposal options, using direct or contract labour. Incineration was also evaluated.

45. Numerous options were initially considered but some of these were dismissed as impracticable and/or environmentally unacceptable eg the disposal of unthickened digested sludge to agricultural land by tanker. This option was dismissed because of the number of vehicle movements and the large distances between agricultural land and the treatment works.

46. Other options considered were various reclamation processes whereby usable products are extracted from the sludge or the sludge is used in the production of some saleable item. None of the processes considered were either sufficiently technically proven, economically viable or operationally secure.

47. After careful consideration of the technical practicalities and the effects on the operational security of the options a final list of twelve variations on sea disposal and five variations of land disposal plus incineration were assessed financially.

48. Included in these options were four thickening options belt press, plate press, centrifuge and Aercon^R. Sea disposal was considered using direct labour and contract labour, operating either the existing fleet or a replacement fleet.

49. The land disposal options are limited to landfill as the only practicable and operationally secure land based option. This created problems with the financial assessment as existing costs of dumping at existing landfill sites would almost certainly increase with increased demand. Alternative landfill sites in Essex and Kent would need to be acquired and planning permission obtained. However despite these problems costs have been ascribed to these options.

50. Incineration was included as this is the only route left if land and sea disposal routes are ruled out for whatever reason.

51. The solids concentration of output sludges ranges from the current 2% without ·thickening, through 5%-7% with various forms of thickening to 19%-30% with various forms of dewatering.

52. All the options considered have some environmental effects, some more than others. For instance, thickening of sludge generally will change the dispersal characteristics at sea; the use of polyelectrolytes with belt presses, plate presses and centrifuges could have repercussions on disposal; land tipping involves substantial vehicle movements and possible river/aquifer pollution; incineration will create air pollution problems and will involve the destruction of a potential resource.

53. The costs associated with the various options (Table 2) were assessed over 30 years and also over 10 years. The latter was carried out in order to obtain a financial perspective of improving the sea disposal option with subsequent termination towards the end of the 1990's, as per the EC Directive.

54. It can readily be seen from the table that the sea disposal schemes with thickening are the cheapest options (10 year NPV costs range from £23 m to £28 m). There is then a mixed group of schemes, including the current operation, which cost in the range of £32 m-36 m (10 year NPV). Land based options are the next group of schemes costing in the range of £40 m-£47 m (10 year NPV). Lastly and not surprisingly incineration was the most expensive option costing £53 m.

55. The effect of calculating 30 year NPVs had virtually no effect on the order of schemes and the general grouping.

CONCLUSIONS
56. In running an effective sludge disposal operation, Thames seeks to obtain the best value for money consistent with safeguarding the environment and maintaining security of operation.

57. The disposal of sewage sludge constitutes only a minor part (some 5%) of the total pollution load to the North Sea and the present practice has not been shown to be detrimental to the marine environment.

Table 2. Sludge disposal options

Scheme No	Scheme Description	Solids content %	Initial Capital £m	Operating Costs £m pa	10 year NPV £m	30 year NPV £m
1	Thames existing operation four ships	2	-	4.6	36	76
2	Contractor operating the existing fleet	2	4.0	3.1	29	57
3	Contractor operating four new ships	2	23.8	2.0	32	56
4	Belt press Thames operation two existing ships	7	7.1	3.7	34	69
5	Belt press Contractor operating two existing ships	7	9.1	2.7	28	56
6	Belt press Contractor operation one new ship	7	14.0	2.1	27	51
7	Centrifuge Thames operation two existing ships	7	5.7	3.7	34	68
8	Centrifuge Contractor operation two existing ships	7	7.7	2.7	28	55
9	Centrifuge Contractor operation one new ship	7	12.6	2.2	25	50
10	Aercon^R Thames operation two existing ships	5	6.0	2.9	28	61
11	Aercon^R Contractor operation three existing ships	5	8.1	2.2	27	54
12	Aercon^R Contractor operation two new ships	5	17.9	1.3	23	48
13	Centrifuge cake to Bedfordshire by road	19	7.6	5.0	45	86
14	Centrifuge cake to Essex/Kent by road	19	7.6	4.5	41	79
15	Plate press cake to Bedfordshire by road	30	15.8	4.0	44	82
16	Plate press cake to Essex/Kent by road	30	15.8	3.6	40	75
17	Plate press cake to Bedfordshire by rail	30	15.8	4.5	47	89
18	Incineration	30	41.1	3.2	53	101

58. Comparison of the land based alternatives with sea disposal shows potential for localised detrimental effects on the environment eg pollution from leachates, excessive vehicle movements in urban areas, air pollution etc. Consequently when compared to the land based alternatives, the sea disposal of sewage sludge from London is the Best Practicable Environmental Option for Thames Water.

59. Sea disposal is the most economic disposal route for sewage sludge but Thames is continuously seeking to improve its sludge disposal operation and the introduction of sludge thickening will reduce the cost of the present operation.

60. However, Thames' research is not only orientated towards improving economics, it is committed to investigating all aspects of sludge disposal including environmental impact. If the existing or any future investigational programme were to indicate that sea disposal was not the BPEO then Thames would respond and the current approach will facilitate any such required response.

61. Thames opposes the draft EC Directive but alternatives to sea disposal will be required if the Directive is approved. Therefore alternatives to sea disposal from London have been identified and are limited to landfill and incineration. The option of conventional land disposal is both environmentally and economically unacceptable.

ACKNOWLEDGEMENTS
62. The authors would like to thank the many colleagues who have contributed to this paper, in particular, the Operations Directorate and the management of Central Division. In addition, the authors would like to thank Mr W R Harper, Managing Director for his permission to publish this paper.

REFERENCES
1. House of Lords Select Committee 17th Report 1986

2. TR 182 WRc August 1982

3. TR 205 WRc July 1984

4. J Orr WRC June 1985 Unpublished Report No 989-M

5. TR 62 MAFF 1981

6. TR 63 MAFF 1982

Discussion on Papers 10-13

MR MIDMER and MR BROWN, Paper 11

There are five towns in the Thanet area with a population of
180 000: large in British terms for seaside towns. The area
embraces ten miles on the north coast and seven on the east.
Before 1973 the three district councils considered the report
of a consulting engineer. The first choice of a joint
outfall at Broadstairs was rejected, and a revised scheme
agreed for a joint sewage treatment works at Richborough. In
the late 1970s a reservoir enquiry drew attention to the
possible use of effluent in the River Stour.

This background forced the careful consideration of
alternatives to the present sites where outfalls discharge.
There are twin pipes at Margate built in the chalk cliff
(Fig. 1), and Broadstairs (Fig. 2) is rather similar with
cliff protection. At Ramsgate the harbour is growing and
getting closer to the outfall (Fig. 3) and nearby Pegwell Bay
has a long sand foreshore.

The Paper sets out the way alternatives were considered.
Surveys to find possible outfall sites were vital and they
were found for each town at increasing distances, the longest
of which was at a distance of 5 km from Ramsgate. Table 5 in
the Paper shows the results and highlights the transfer costs
in the single site options, aggravated by lifting to cliff
tops.

The serious sewage treatment works options involve
discharge into the River Stour. This must be well upstream
to avoid bacteria on beaches at the mouth, with discharge to
the inside of the loop, rather than outside as was originally
considered. The effluent needs to be a high standard to
avoid changing the quality of the Stour; the most important
component would be ammonia, with a likely consent level of 10
mg/l.

The resource benefits are set out in Table 2 in the Paper.
These result from additional abstraction upstream and are
derived from a resource model used for the Kent rivers supply
area. The lower benefit shown with the combined treatment
works, compared with the other options, is due to the cost of
getting the water into supply. Assessments of this kind rely
on accurate estimates of both capital and running costs. As

Fig. 1. Existing headworks at Margate

Fig. 2. Existing headworks at Broadstairs

218

Fig. 3. Ramsgate harbour

a capital project an outfall is a major undertaking.
Inevitably, the need to get public acceptance of the project
influenced the proposals put forward.

We have touched on the work which was done to inform
councils and the general public. This enabled the authority
to get COPA consents for both outfalls without significant
objection. Separate schemes for the three towns were
instrumental in gaining support. Dealing with one's own muck
is more acceptable.

MR B. A. O. HEWETT, Southern Water
Papers 10-13 on economic considerations come from four water
authorities in the UK. All of them, including Thames, who
deal only with sludge disposal, consider marine treatment to
be significantly more economic than all other options
containing inland treatment works. In the three Papers
dealing with long sea outfalls it was most interesting to see
the different approaches adopted to arrive at this
conclusion.

Welsh Water and Southern Water adopted a strategic approach
to the problems of sewage treatment and disposal over large
areas covering several major sites. Southern Water even
incorporated the option of using the effluent from a proposed
inland sewage treatment works at Ramsgate as an indirect
water resource, enabling abstraction for water supply

purposes to be increased upstream of the works discharge point on the River Stour. This broad approach has also been adopted where appropriate by the other authorities, aided no doubt by the power of computers to examine the 'what if' situation.

It was the Welsh approach, however, which contained some original thinking in the way they attempted to correlate the results of the costs survey they undertook, in the form of a function which measured cost per unit length against diameter of outfalls with an appropriate allowance for mobilization costs. Fig. 2 of Paper 10 shows the results in graphical form. I was intrigued to see on the bottom right-hand side of the graph that the Authors used tender prices and contractors' budget quotes. It did not take a magnifying glass to see that there were differences between the two, as so many of us have found when we have had to explain some embarrassing figures to our Boards! I suggest actual end of construction costs should also be included. Perhaps the Authors could elaborate on the data and its value.

Wessex Water (Paper 12) drew attention to the fact that on large schemes, which all tend to be unique in some respects, 'appropriate risk factors should be considered if a proper appreciation of the likely capital costs is required'. This applies particularly when tunnelling is involved, either under sea or as part of the transfer scheme. Sensitivity analysis clearly plays a part although it is not mentioned specifically in terms of possible additional construction costs.

Perhaps the Authors of all three outfall papers might like to comment on this aspect in the light of their post-project experience. Would an economic appraisal after the events in question have produced the same answers?

Wessex Water carried out an interesting review of operational costs of their headworks after three years' experience which, although 10-20% higher than expected, still produced figures very much less than an equivalent inland works. It was noticeable that labour, transport and building maintenance still accounted for over a third of operating expenditure in spite of all the automated equipment. However, the capital costs, particularly the phasing of expenditure, are the dominant factor. The latter is a very important factor with a tight capital programme and equally pressing demands for improvement elsewhere.

Welsh Water bring out this point of the phasing of financial expenditure with the statement that a target figure of £3 million was established as the maximum that would be available each year without inhibiting other priority capital investment. Would a different figure have produced different solutions?

I know that Southern Water have a priority allocation system which looks at each capital bid across all functions on a points basis. The first run of the authority's capital programme on the computer is usually on this basis, although

adjustments are made to take account of other factors which
are not easily quantifiable. However, I saw no overt mention
of the cost-benefit approach, or perhaps more accurately the
cost-disbenefit approach. It was there in the recognition of
water resources benefits by Southern Water and the
disbenefits of an adverse impact on the roadworks programme
by Welsh Water. Should we be moving towards the more formal
approach of an environmental impact analysis available to the
public? They would then see the scale of economic and other
benefits and disbenefits more clearly, particularly on such
problems as the cost of disruption to the business community
of large-scale works and other environmentally sensitive
issues. The wider economic impact on the community must be
seen to be considered.

Perhaps we keep these matters too much to ourselves. We
have only ourselves to blame if the public rely on instinct
and prejudice to reject our proposals for marine treatment if
they are not shown all the considerations, economic or
otherwise, which have been taken into account. The increases
in capital expenditure needed to respond to an embargo on
marine treatment would surely justify the attempt.

DR T. J. LACK, WRc Environment

A great deal of emphasis has been laid on designing outfalls
to comply with the EC bathing waters directive. The
assumption seems to be that if the bacterial standards in the
directive are complied with, then public health is assured
and capital expenditure is justified. The directive also
refers to a standard of zero enteric viruses per ten litres.
I would like to ask Dr Pattinson and Mr Jones if any outfall
has been (or indeed can be) designed to this standard. It
appears to me that outfalls may have to be so long that they
become economically unattractive and/or some form of
disinfection may need to be incorporated into the headworks
processes.

I would also like to ask Dr Pattinson or any other Author
if they are satisfied with current knowledge of the survival
of viruses in sea water. What are the implications for the
future design and costs of outfalls if current public
awareness of viral illnesses leads to more stringent
application of viral standards?

MR M. KING, Land & Marine Engineering Ltd

Southern Water have taken positive action to clean up the
bathing waters along their extensive and prestigious
coastline. They saw the clear advantages of marine treatment
and built, I think, something like five major outfalls in as
many years - a truly magnificent achievement. As is clear
from Paper 11, each case has been investigated in great
detail, the options studied in depth with open minds. The
preferred solution - from the outset, it seems to me, was
marine treatment preceded by normal headworks screening and
maceration. But at Ryde on the Isle of Wight, pre-treatment,

inland, by a novel method was seen to be the right solution - followed by a relatively short outfall. And I have been very interested, for some time in the options being studied for Ramsgate and Sandwich. Not every location - or every coastal sewage disposal problem - is capable, of course, of a common solution. Mr Midmer and Mr Brown have demonstrated the reasons why treatment works were the right option for Ramsgate and Sandwich.

The public must have gained confidence from the open and flexible approach which Southern Water has shown and in their readiness to adopt not necessarily the favoured solution, but the right solution to a particular problem. Politically this must have been good. I like, too, their policy of taking the public into their confidence at a very early stage. Obviously this approach has paid off and lessons will have been learned by others. A high standard of effluent will be expected - or rather demanded - from the Richborough site which will serve Ramsgate. Has a decision or process selection been made and if so what is it?

MR W. HALCROW, Forth River Purification Board
The Authors of Paper 10 rightly draw attenion to the importance of flow estimation in the design of effective outfalls. I am, however, surprised that they do not mention storm water, particularly in a high rainfall area which will have a complex sewerage system with many potential points for storm overflows.

There can be no doubt that any disposal scheme must make proper provision for storm water. In terms of compliance with quality criteria unsatisfactory storm overflows can negate all the work done to dispose of the foul flows by causing both transient and possibly more enduring pollution of bathing areas. What provision has been made for storm water disposal in the scheme? More generally, what realistic options do the Authors see for the treatment of storm water from sea disposal schemes in high amenity areas?

MR J. A. WAKEFIELD, Coastal Anti-Pollution League
Does Mr Jones expect to receive an EEC grant for the work at Llandudno and Colwyn Bay?

MR D. F. H. PHAROAH, W. S. Atkins & Partners
Many contributors have compared the options of long sea outfalls and inland treatment works, concluding that for coastal areas, in most cases, both from treatment and economic considerations, long sea outfalls are preferable. The nature of the construction of a sea outfall demands that the designer should aim for a very long design or asset life, say, approaching one hundred years. The use of long asset lives implies that change will not occur or, where it does occur, it will not affect the use of the outfall. Predicting the future is notoriously difficult. However, changes can occur from new legal constraints (EEC), medical discoveries,

increasing environmental knowledge, population shifts, or differing industrial discharges.

Are designers seeking to establish factors for the probability of changes and using them to carry out asset life risk appraisals? Where such techniques are used they enable the designer, when considering options, to make weighted judgements and not to rely on simple economic or ·scientific results to reach their conclusions.

DR PATTINSON and MR JONES, Paper 10

In reply to Dr Lack, our present level of knowledge on the behaviour and fate of viruses in sea water is insufficient to allow us to design outfalls which will reliably achieve the very stringent virus standards specified in the EC bathing water directive. This is of great concern as there is much talk of building outfalls which achieve the requirements of the EC bathing water directive when in fact they will only reliably achieve the E.coli and coliform standards. Such outfalls may fail the overall requirements of the directive by not meeting the virus, salmonellae and/or faecal streptococci standards. An extensive monitoring programme in Welsh bathing waters in 1986 indicated that viruses were ubiquitous and many beaches which reliably passed the bacterial standards of the directive failed to meet the virus standard. Furthermore there was no evidence of any correlation between bacterial levels and virus concentrations. This is not to say that properly designed long sea outfalls do not have a dramatic effect on virus levels in bathing waters. For example a new outfall commissioned by Welsh Water at Tenby in 1986 immediately achieved the EC E.coli and coliform standards and has, over the intervening period, produced a tenfold reduction in virus levels with a continuing downward trend. However, at the present time this beach still fails to achieve consistently the virus standard in the directive and it is salutary to reflect on the timescale over which this reduction is taking place and our inability to predict the final outcome.

This situation suggests a need to pursue vigorously the present investigations at Tenby into the fate of viruses in sea water so that the industry can design sea outfalls to achieve virus standards consistently. This work must clearly be supplemented by a major epidemiological study to establish the significance of existing virus standards with respect to the protection of public health. These studies should not be directed solely at quantifying the risk of serious disease. There is increasing concern over the incidence of minor ailments arising from bathing as highlighted by the 10th report of the Royal Commission on Environmental Pollution and it is in this area that attention must be directed if marine treatment is to remain an effective and justifiable disposal option in the eyes of the public. Welsh Water does not consider disinfection to be an environmentally acceptable solution to the problem of viruses in effluents discharged to

223

the marine environment in view of the production of persistent chlorinated compounds which can accumulate to significant levels in indigenous organisms.

In reply to Mr Halcrow, the impact of storm overflows, their location and frequency of operation, is crucial to the achievement of satisfactory bathing water quality and is probably the major outstanding technical issue which must be addressed if the industry is to design effective sea disposal schemes to achieve the requirements of the EC bathing water directive. Unsatisfactory storm overflows can give rise to two problems about bathing water quality: firstly aesthetic contamination with sewage debris and secondly bacterial contamination which could cause failure of the EC bacterial standard if the frequency and duration of the storm discharge is sufficiently high.

In the Llandudno scheme existing discharges will be used as SWOs and their impact on bathing water quality will be minimized in two ways. Firstly, for those SWOs in the immediate vicinity of beaches, the weir settings will be sufficient to ensure that the overflows operate for less than 5% of the time during the bathing season so that bathing water bacterial quality will achieve the 95% standard specified in the directive. Overflows will be allowed to operate more frequently elsewhere in the scheme so that unacceptably large flows are not carried forward to the outfall. Secondly, where possible overflows will be physically extended into the main channel to maximize available dilution and reduce the incidence of high bacterial concentrations. This is a pragmatic approach adopted by the authority and highlights the urgent need to develop predictive techniques which will allow confident prediction of the duration and impact of (interacting) storm events on bathing water quality and assist in optimizing overflow settings. The level of preliminary treatment on storm overflows to high amenity areas must also be critically examined if sea disposal schemes are to achieve the high aesthetic criteria demanded for bathing waters and the impact of this on the economics of sea disposal must be reviewed. It may well prove necessary to provide fine screening on overflows directly into amenity areas and in these situations it may be more appropriate to locate the overflows outside amenity areas to allow lower levels of treatment.

In reply to Mr Hewett, Welsh Water's main concern at the time of formulation of strategy was to establish a cost function which could be used to forecast quickly and reasonably accurately the cost of a whole series of possible outfalls. To achieve statistical accuracy, as many costs as possible were built into the model, contactor's budget quotes included. It is true that historically there has been some difficulty in correlating budget quotes with actual tender prices. A further analysis of the costs used in determining the model shown in Fig. 2 of the Paper, but excluding budget quotes, gives a revised function as follows tender price (£) = 1.2 diameter (mm) (length [m] + 500)

This gives tender prices 20% higher than the equation
included in the proposal, but it must be borne in mind that
the considerable reduction in the sample number will give a
corresponding reduction in accuracy.

Unfortunately Welsh Water have only agreed final costs on
one outfall scheme to date and the amount of information
available from other sources was similarly limited. Because
of the lack of data on the tender price/final cost ratio, it
was not considered practicable to formulate a costing
function at this stage, although it is hoped to do so as more
information becomes available. The indications are, however,
that in some cases at least, the final cost may be
significantly in excess of the tender price. If this proves
to be a particular feature of outfall schemes, then it should
be recognized as soon as possible and built into the initial
economic appraisal. Post-project investment appraisals are
useful in this respect, not only to discover whether the
correct investment decision was made, but more significantly
to provide up-to-date data for refining the initial appraisal
method.

The availability of capital did indeed prove a considerable
constraint in the formulation of strategy. The approach
taken by Welsh Water during the investment appraisal was to
determine the most cost-effective practicable scheme using
the NPV and sensitivity analysis described in the Paper, and
then to determine whether it was possible to phase that
particular scheme within the capital constraints. It was
fortunate that it proved to be possible to accommodate the
favoured scheme within these constraints, although a higher
level of capital availability would have allowed much more
flexibility in programming the works.

With regard to the question on the amount of information
made available to the public, I personally feel that this is
an area where the water industry has at times failed in the
past and where we are learning only slowly. We are used to
arguing our investment decisions to boards and senior
management, but sometimes forget that, particularly where
outfalls, and a tourist-orientated economy are involved, the
public are deeply concerned over the way several million
pounds may be invested to solve a problem which they perceive
as affecting their livelihood. I feel that we should give
much more of an emphasis to selling our proposals to the
public and that will involve identifying and addressing those
problems with which the public are most concerned. I suspect
that the strategy to be adopted to implement this may prove
to be somewhat different in each case and could involve any
of Mr Hewett's suggestions such as cost-benefit and
environmental impact analysis, linked to a positive
information campaign. This of course takes time, effort and
money, all of which I think worthwhile if the general
agreement and consent of the public can be achieved.

In reply to Mr Wakefield, the main criteria whereby the EEC determine whether a scheme can be grant-aided by the European Regional Development Fund is on the basis of geographical location within a designated development area. None of the Colwyn and Aberconwy area is within a designated area, so it is unlikely that any grant aid will be forthcoming from the EEC. Welsh Water always apply for whatever grant aid they consider may be available, and an EEC contribution was in fact received for the outfall recently constructed at Tenby.

In reply to Mr Pharoah, the approach taken by Welsh Water to possible changes in flows was to carry out sensitivity analysis at various flow levels to ensure that the chosen strategy held good for the whole range of predicted flow changes. With regard to industrial discharges it is usual to assume that any new discharge will be treated at source. Factors for the probability of changes due to new legal constraints, medical discoveries and increasing environmental knowledge are very difficult to determine and build into economic analysis. It is possible, however, that should any of these changes come about within the asset life of any scheme, then they could be catered for by changes to the level of pre-treatment applied to the outfalls.

MR MIDMER and MR BROWN, Paper 11

With regard to Mr Hewett's comments on economic appraisals, it is essential to be able to assess qualitatively the adequacy of all estimates. Southern Water makes use of two checks. Firstly, a sensitivity analysis is carried out to establish by how much the captial and operating costs must change before the preferred scheme is no longer the most cost effective. Secondly, the present value is established at discount rates of 2.5% and 7.5%, that is either side of the test discount rate of 5%, and the preferred scheme is required to remain so over this range; if this is not the case then the various options are examined in greater detail. The procedures appear to work satisfactorily and post-project experience in Thanet has confirmed the conclusions reached in the feasibility study. However, it must be accepted that all forecasts are subject to unknown error and the use of the risk factor approach would be helpful in providing further guidance as to the robustness of the chosen solution.

Mr Hewett's comments on the wider approach to appraisal are interesting. Certainly there is a tendency to concentrate on the development of cost-effective solutions and, although the benefits and disbenefits associated with a scheme are examined, the assessment may well be somewhat subjective. Bearing in mind the difficulty of awarding a cost to a benefit or disbenefit this is perhaps understandable; after all, what is the value of a clean bathing water? However, it would be in the interests of the industry and the public to demonstrate that the wider impact on the community is being considered and the environmental impact analysis approach would be appropriate.

Dr Lack's question reflects the growing interest in viral standards. In the past the industry has tended to concentrate, for a variety of reasons, on bacteriological standards and the present knowledge of the survival of viruses in the sea and therefore the implications for design are not satisfactory. This is certainly one of the areas within the field of marine disposal technology where further work is required, and in this respect the work by WRc and Welsh Water is very important. It remains to be seen whether improved understanding of viruses leads to changes in outfall practice but it is encouraging to note the results at Tenby (Paper 14) which indicate significant reductions in enteroviruses after commissioning the outfall. The small number of viruses still being recorded are thought to be associated with other discharges and, significantly, a storm overflow. The design and operation of storm overflows is clearly another area where further detailed work is required.

With regard to Mr King's question, the treatment systems considered have been greatly influenced by the need to achieve high ammonia removal. The feasibility study examined percolating filter, traditional activated sludge and a range of oxidation ditch systems, and the latter proved to be the cheapest overall. The initial findings are currently being re-examined.

Mr Pharoah is concerned with assessing the probability of changes occurring over the life of a scheme. It is important when developing and appraising a scheme to take into account situations and events which can reasonably be expected to occur. However, it is also important not to be too speculative. Changes in population and water usage, for example, will be assessed and, if considered necessary, the effects of changes in the assumptions can be incorporated in a risk factor analysis. Similarly, pending legislative changes can be incorporated. However, it would be unreasonable to speculate on future developments in areas such as medicine and environmental knowledge and attempt to take these speculative changes into account.

MR TYLER, Paper 12

Mr Hewett enquired about sensitivity analysis and post-construction appraisal. Both techniques are employed routinely in Wessex to try to ensure that the best overall option is selected in the first instance, and to see what lessons, if any, can be gained from a thorough examination of post-construction costs and option performance.

In the case of Weymouth and Portland, the sensitivity analysis took the form of a range of possible cost variations reflecting both the detail of the estimate and the type of construction involved. These costs are listed in Table 1 of the Paper. As Table 2 shows, the range of uncertainty for the marine treatment option was significantly underestimated for reasons which have been discussed in the Paper. One of the lessons from post-construction consideration of the

project is that the sensitivity of marine treatment and
tunnel construction works to cost variation should be
recognized and allowed for adequately in the appraisal
process. Table 2 also shows that the final costs of the
marine treatment scheme as constructed, although higher than
predicted in 1979, are still less than any of the other
options considered at that time. Since it is virtually
certain that other options would have experienced cost
increases to varying degrees, there is no doubt that the
option selected was the correct economic choice. The Paper
also considers operating costs in some detail and confirms,
as expected, that actual costs are only slightly higher than
forecast, and considerably less than those for an equivalent
sewage treatment works.

The most important post-construction appraisal exercise of
all, however, is to determine whether the investment has been
worthwhile in achieving the objectives originally set. It is
pleasing to report that considerable shoreline monitoring has
demonstrated the efficiency of the marine treatment scheme in
ensuring that all coastal areas can meet Eurobeach standards,
without any deleterious effects on the marine environment.

Regarding environmental impact analysis, it is a matter for
each authority to determine the degree of formality and the
public accessibility to the results. What is undeniable is
that there will continue to be opposition to marine treatment
ranging from uninformed prejudice to the more articulate, and
no scheme should be proposed and presented to the public
without full consideration of the environmental effects.

In reply to Mr Pharoah, most economic appraisals consider
that the full asset life can be utilized unless there are
particular reasons why replacement may be required before
then. Mr Pharoah's points are interesting, and it is
certainly true that the future even 20-30 years hence is
difficult to predict. Although the possibility of changes in
legislation, etc., which may affect sea outfalls is a matter
which should always be borne in mind, there are similar
difficulties in predicting the long-term acceptability or
suitability for inland treatment options also. These matters
should be assessed properly in the appraisal stage, and
considered judgements made, but it would be difficult to try
to quantify these effects, and any such attempts are likely
to be unreliable.

MR GREEN and DR HANBURY, Paper 13
There some relevant points to make on four related topics:
the recent technical conference on the North Sea organized by
WRc, the draft EC directive on the dumping of wastes at sea,
the disposal of sewage sludge in particular, and Thames
Water's stance towards sludge disposal and our present
operation.

The North Sea Conference covered virtually all the
environmental aspects of the North Sea. There were several
main conclusions.

(a)   The two main sources of pollution to the North Sea are
      the river estuaries (notably the Rhine, Meuse and Elbe
      which are grossly polluted compared with the Thames),
      and atmospheric pollution. The Paper puts sludge
      dumping at less than 5% of the total pollution load to
      the North Sea. It is now considered that the main
      problem in rivers is diffuse pollution sources rather
      than point sources.
(b)   The main areas affected are coastal waters, in
      particular the Dutch and Danish coasts, but there are
      also isolated areas showing effects of pollution such
      as the Dogger Bank and the German Bight.
(c)   The mathematical modelling work that has been done
      shows the North Sea to be divided into two halves.
      The northern half is deep, well mixed and has a rapid
      turnover of water with input from the Atlantic. The
      southern half is shallow, has an anti-clockwise
      circular flow pattern and a low turnover of water.
(d)   Finally, it was generally agreed that in absolute
      terms little is known about the very complex system of
      the North Sea.

The draft EC directive on the dumping of wastes at sea is
one of the examples where there is a difference of opinion
between the UK and the Continent, as it is based on the
anticipation principle that all discharges are undesirable
and harmful and should therefore be reduced or stopped
altogether. Bearing in mind that it has been identified that
rivers and the atmosphere are the main sources of pollution,
it seems unproductive to concentrate efforts on dumping
restrictions. The directive also takes a very blinkered
approach in that it calls for restrictions on disposal to sea
and recommends that other routes are used but makes no
attempt to consider the consequences or indeed allow for
comparisons between routes. Along the same line the
directive does not consider any of the technical, economic or
practical factors involved in introducing reductions or bans.
This point was recognized as important by the House of Lords
Select Committee.

Sewage sludge is only one of the substances covered by the
directive but seems such a small proportion of the total
pollution load as not to warrant inclusion. To illustrate
that point Table 1 gives some figures for metals discharged
into the North Sea taken from the 10th report of the Royal
Commission on Environmental Pollution. Similar orders of
magnitude are observed with organic carbon figures.

Indeed, there is little scientific evidence to show that
the low level of contamination by sewage sludge causes any
genuine pollution. This distinction between contamination
and pollution is another example of differing views of the UK
and the Continent. The last and probably the most important
example of this difference in approach is that the UK
considers the concept of best practicable environmental

Table 1.   Sources of pollution - estimated contaminant
loads to the North Sea (tonnes/year)

Source	Cu	Zn	Pb	Hg	Cd	Ni
Rivers	1950	13840	2430	30	280	1620
Atmosphere	3450	10750	4500	6	530	1650
Direct discharges	310	5570	560	6	20	310
Dumping of sewage sludge	120	390	110	3	10	60
Dumping of other wastes	190	520	240	<0.4	<0.4	100

option to be the guiding principle where sewage sludge is
concerned.

With regard to Thames Water's stance concerning its present
sea disposal operation, the entire operation is carefully
managed and controlled to ensure that everything is done in
accordance with the MAFF licence. Thames' approach is to be
as aware as possible of the effects of the operation and if
necessary to modify the operation where appropriate to
counteract any deleterious consequences. This approach has
been put into practice in the form of the study currently
being undertaken jointly with WRc. Initial results of the
study have recently been published and the conclusions to
date show the dumping site to be highly dispersive; there is
no clear evidence of detrimental effect on macrobenthic
community and that the incidence of fish disease is low and
similar to control areas.

Thames Water is continuously seeking to improve its
operation both on the effectiveness side, for example by the
trade effluent control highlighted in the Paper, and also on
the efficiency side by the type of analysis and investigation
into possible alternatives also described in the Paper.
Thames' overall approach is to have the best disposal
operation possible taking into account all factors and not
simply to adopt the least cost option.

14. Monitoring for environmental impact

V. A. COOPER and T. J. LACK, WRc Environment, Medmenham

SYNOPSIS. Increasing environmental awareness means that the UK water industry must be in the forefront in assessing the effect of marine disposal of sewage and sludge if they wish to continue to benefit from this outlet.

The Water Research centre is currently involved in research into the effects of sewage and sludge disposal on the environment. The broad based studies include computer modelling for impact prediction, tracer studies, viral surveillance techniques, ecological impact and the monitoring of sub-lethal effects using field and laboratory bioassays. In addition there are studies aimed at producing guidelines for hydraulic and structural design of sea outfalls and quantifying the effectiveness of head works processes.

Adverse effects detected in the field have been minimal, even using highly sensitive techniques. The importance of broad-based studies in such areas of minimal ecological impact is stressed.

INTRODUCTION
   1.   The UK Water Industry is heavily dependent upon the use of the marine environment for the treatment and disposal of waste. Approximately half of all sewage produced in Scotland and Wales and around one-quarter in England is discharged through outfalls to the sea. This represents the waste of approximately 16 million people. In addition, about 30% of the sludge arising from sewage treatment works is disposed of to sea at licensed disposal grounds.
   2.   Increasing environmental awareness has led to resistance to marine disposal both from environmental pressure groups and from other European countries. If the UK wishes to continue to benefit from the marine disposal outlets, it needs to be seen to be adopting a responsible approach with regard to the environmental implications. This will require the water industry to be at the forefront in assessing the effects of current practice. This paper reviews the research programme put in hand by the Water Research centre. Attention is focussed firstly on the prediction of sewage and sludge dispersion and bacterial mortality in the sea. The broader environmental effects are

then reviewed, including the use of novel techniques to assess environmental impact. The aims of the case studies are twofold: firstly to demonstrate that the sea has the capacity to both assimilate and treat sewage and sludge, provided that the disposal methods and locations are correctly chosen; secondly, to develop and assess new techniques in monitoring as part of broad-based studies.

BACTERIA AND VIRUSES

3. <u>Viruses</u>. Viruses in sea water are an increasing public health concern. Although total numbers of viruses in sewage are low, with a correspondingly low health risk, there is evidence that viruses can survive for prolonged periods in sea water, which is of concern to designers of long sea outfalls. For example, enteroviruses have been reported (1) to survive for 14 days in clean sea water and 35-40 days in the presence of sewage solids and other suspended particles. Viruses tend to clump together in the water attached to particles and to each other, and have also been reported to attach themselves within folds of shellfish tissue, where they can survive for several months.

4. WRc is currently involved with a study on the distribution and survival of rotaviruses in the environment, contracted out to the Welsh Water Virology Unit. Rotavirus is associated with clinical forms of gastroenteritis and is responsible for approximately half the cases of infantile diarrhoea that require hospital treatment worldwide. Enteric viruses can be excreted in concentrations of more than 10 per gram of faeces and more than 10 per litre have been detected in raw sewage. Viruses are naturally embedded in faecal materials and remain associated with the solids even after the sewage is broken up. During primary sedimentation about 60% of the virus load settles with the sewage solids into primary sludge. However, rotaviruses bind less efficiently to activated sludge solids and may not be as effectively removed by treatment as are enteroviruses. This may result in the selective discharge of rotavirus into the environment. The Welsh Water Virology Unit is developing a method for routine measurement of rotavirus in marine environmental samples. Enterovirus and Rotavirus are being measured in water, sediments (including a bathing beach) and shellfish tissue from a case-study site in Wales.

5. <u>Bacteria</u>. The EC Directives on the quality of bathing water (2) and the quality of shellfish waters (3) both set standards for the acceptable numbers of faecal coliform bacteria. The bathing waters Directive sets an imperative value of 2000 faecal coliforms per 100 ml. The shellfish waters Directive sets standards for the fluid within the shellfish, but not for the water in which they live, which makes it impossible to design to. The Ministry of Agriculture Fisheries and Food (MAFF) should be consulted for advice on acceptable bacterial concentrations in shellfish waters, but suffice it to say that considerable reductions in bacterial numbers are necessary between an outfall and either bathing waters or shellfish beds.

6. Mathematical modelling. A suite of models has been
developed to simulate, in a quantitative way, the dispersion
of sewage and other potentially polluting materials.
Dispersion from single or multiple point sources into tidal
rivers, estuaries or open coastal waters can be modelled and
account is taken of local hydrodynamic characteristics, shore
topography and bed properties. The sewage dispersion model
is proving to be an invaluable tool by assisting in the
design, location and operation of sewage outfalls. Any
number of design, location and operational options can be
modelled under a wide range of physical and meteorological
conditions thereby bringing about considerable savings in
time and money.
7. Tidal currents and water levels are calculated over
tidal cycles using a hydrodynamic model. This is validated
using field data derived from float/drogue tracking and
bacterial spore dispersion studies.
8. The hydrodynamic model is interfaced with a random
walk, bacterial dispersion model which takes account of
wind-induced movements by superimposing a parabolic vertical
profile on top of the depth averaged current profile. After
running the dispersion model over several tidal cycles,
estimates of bacterial numbers in small cells over the sea
area are stored at hourly intervals over a whole day and the
display programme then presents the results for any discharge
options in colour graphic form. Differing decay factors and
polluting loads can also be modelled.
9. The model has been widely applied for simulating the
dispersion of faecal coliform bacteria in sewage discharges
with particular regard to predicting concentrations of
bacteria in the sea and at the beach, to ensure compliance
with the EC Directive on Bathing Water.
10. Case Study - Distribution of Bacteria and Viruses The
main case study site is at Tenby (Welsh Water), a popular
holiday resort in Dyfed. The original sewage disposal system
at Tenby was constructed in the 1960's and had become
inadequate for the inflow during summer months when the
population is increased considerably by the influx of
tourists.
11. Effluent disposal was through a 6 foot diameter brick
Sewer connecting with the River Ritec at a ponstock chamber.
Subsequently both shared a cast iron pipeline to a short sea
outfall just below low water mark on the South beach. At low
water on spring tides the pipe end was often visible and both
sewer and pipe were in a bad state of repair. The discharge
was coloured and visible at all states of tide. The bathing
beach waters were contaminated by sewage solids and high
levels of coliform bacteria and enteroviruses, which
frequently exceeded the standards set by the EC Bathing
Waters Directive.
12. Although improved management of effluent release
improved bacterial standards from 1978, as shown in Table 1,
the required 95% compliance was not attained. A new, long
sea outfall (2.7 km) was therefore commissioned in July 1985,
together with a new pre-treatment works where the sewage is

macerated to pass through a 6.0 mm screen. Table 1 shows the improvements in both bacterial and viral standards on the Tenby beaches following commissioning of the outfall. There are still small numbers of enteroviruses found at the beach, but these are thought to have come from the old outfall pipe which still carries the waters of the River Ritec and also some storm overflow from the sewage works.

ENVIRONMENTAL IMPACT

13. WRc is currently involved in two case studies of modern outfalls and a sewage sludge disposal ground. These studies comprise five main aspects: tracer studies to assess the area of impact; distribution of bacteria and viruses (described above); effects on the benthic ecology of the surrounding areas; the accumulation of metals and persistent organics in the sediments; and the sublethal effects on marine organisms. The outfalls being studied are Tenby (Welsh Water) and Weymouth (Wessex Water). Both are modern outfalls discharging into sufficient water depths and currents to allow adequate dispersion. The sludge disposal site is in the outer Thames estuary (Thames Water) and is a dispersive site.

Table 1. Bacterial and viral standards at Tenby

	1977	1978	1979	1980	1981	1982	1983	1984	1985a	1985b
coliforms (% exceedence of standard)	5	43	–	19	19	15	15	25	6	0
enteroviruses (pfu/101)	–	–	–	66	22	40	14	20	19	1.3

notes: pfu = plaque-forming units
       1985a = Jan – July (pre-commissioning)
       1985b = July onwards (post-commissioning)

14. Tracer studies. Sewage and sludge dispersal at the case study sites have been measured using various combinations of methods. For measurement of short-term dispersal, rhodamine dye and radioactive tracers with a short half-life have been used. Rhodamine shows the dispersion in the water column, and a detailed picture can be built up by incorporating the use of multi-spectral scanning techniques. Radiotracers can trace the sewage or sludge both in the water column and in the sediment, but public concern after the release of radioactivity into the environment limits the extent of their use.

15. The short-term tracer most frequently used on these case studies is the distribution of faecal coliform bacteria. Although mortality rates of the bacteria vary and they are not necessarily source-specific, faecal coliforms do give an initial indication of the area of impact of an outfall as in the case of the Tenby outfall, shown in Fig. 1 and in the Thames sludge disposal ground (Fig. 2).

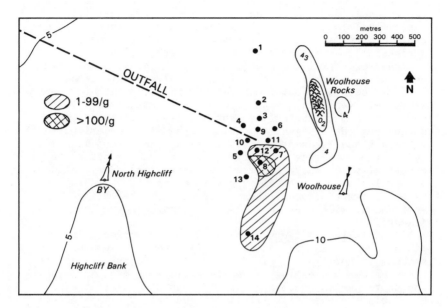

Fig. 1. Tenby outfall, showing distribution of temperature
tolerant coliforms in sediment

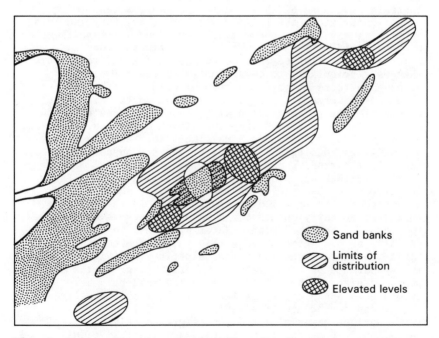

Fig. 2. Thames Estuary, showing distribution of temperature
tolerant coliforms in sediment

16. For medium term tracing of sewage and sewage sludge in sediments, WRc are assessing the use of three compounds: coprostanol, nonylphenol and alpha tocopheryl acetate. coprostanol is a faecal sterol found only in mammals, is not rapidly degraded in the environment and has been used successfully as a tracer by other workers (4,5,6).

Alpha tocopheryl acetate (alpha TA) and nonylphenol have been found in coastal sediments and their occurrence has recently been attributed uniquely to the disposal of sewage effluent and sewage sludge. Nonylphenol is a toxic decomposition product of non-ionic surfactants produced during anaerobic digestion, while alpha TA is a derivative of vitamin E and is used extensively as a stable food additive and in the cosmetics industry. These compounds offer the opportunitý to study the extent of sewage and sewage sludge dispersion at sea, and it is possible that establishing the rates of degradation of these compounds in the marine environment will enhance their use as nonconservative tracers.

17. Ecological effects. The study of the effects of discharges on benthic communities (living in or on the seabed) is not a new one. However, it is important that the study of benthic ecology at outfall sites is continued in conjunction with novel techniques. Benthic invertebrates have relatively sessile lifestyles and cannot escape effects of pollution by swimming to another area. The natural environment in which these organisms live is often hostile and can be completely disrupted by storms or changes in tidal currents. Because of this, benthic organisms tend to be 'opportunistic' in their lifestyle – that is, they can invade an area as soon as it becomes habitable and increase their numbers rapidly. These naturally large fluctuations in numbers make pollution-induced changes difficult to detect. Changes attributable to sewage outfalls are not always adverse in terms of total numbers of animals – the oligochaete worms, in particular, thrive on the high levels of organic matter associated with sewage. The distribution of species may change, resulting in a lower diversity, while the total numbers of individuals may increase. Because of these difficulties, ecological studies at WRc are supported by accumulation studies and bioassays.

18. A further difficulty in interpretation of results is variations in the natural sediment quality, which may be reflected by the faunal distribution. It is therefore important to carry out thorough baseline surveys, such as that carried out by Southern Water at Hythe, where a new outfall is to be commissioned this year. The natural variation in sediment and faunal distribution (Fig.3) gives the appearance of a "plume" effect originating at the proposed site of the outfall. This may well have been interpreted as an effect of the outfall if this baseline survey had not been carried out.

19. On the Thames sludge disposal ground (Fig. 4), the variation in numbers of individuals and species is influenced chiefly by the sediment type. From these results alone, it

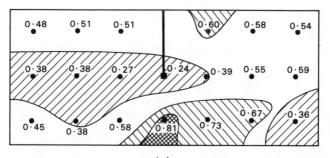

(a)

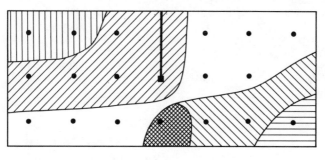

(b)

Fig. 3. Proposed site of Hythe outfall:
(a) percentage organic content of sediment;
(b) cluster groups of faunal affinities

is impossible to say whether sludge disposal activities are
having no effect on the ecology, or whether the effects are
simply masked by the variations in sediment type. It is for
this reason that surveys should be as broad-based as
possible, using techniques such as bioassays. Such work is
in progress on the Thames estuary, but is not reported in
this paper.

20. Probably the most sensitive methods of monitoring the
effects of sewage discharges are sublethal tests. WRc are
monitoring the case-study outfalls and sludge disposal site
both on site and in the laboratory.

21. Under the controlled conditions of the laboratory,
dilutions of sewage and water samples from around the two
case-study outfalls have been used for bioassays on mussel
and oyster larvae. These tests examine the effects on
development of the sensitive early life stages and

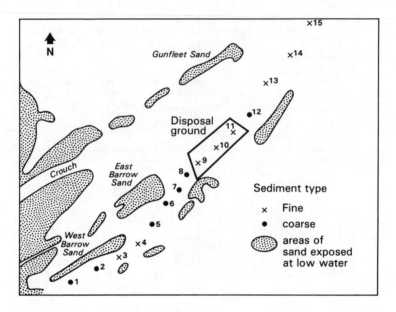

(a)

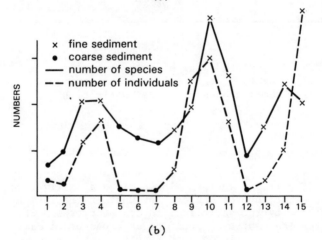

(b)

Fig. 4. Thames benthic survey: (a) sampling sites and sediment type;(b) numbers of invertebrates and species at each site

experiments to date indicate that they are capable of monitoring levels as low as the EQS for ecosystems. Results from Tenby outfall are shown in Figs. 5a (sewage dilutions) and 5b (water samples from around the outfall at sites shown in Fig. 5c). Although the 10% sewage solution showed a reduced mean percentage development of larvae (Fig. 5a) none of the sewage concentrations were significantly different from the control (at the 5% level of significance according to the Scheffe test). It is not surprising then, that the environmental samples (Fig. 5b) showed no significant effect on larvae.

22. Monitoring in the field is achieved by deploying a series of cages of mussels around each outfall. After at least 30 days' exposure, the mussels are removed and taken back to the laboratory where physiological and cytochemical tests are carried out. These tests are being evaluated under contract to the Department of the Environment, and are described in detail in a special volume of Marine Pollution Bulletin (7). In brief, the tests examine the physiological wellbeing of the mussels (scope for growth) and the initiation of detoxification mechanisms which can be measured by lysosomal latency (general stress), a metal stressor test and an organic stressor test. These tests are supported by measurements of body condition and metal accumulation.

23. The Scope for Growth (SFG) technique was first used and reported by Widdows et al (8) in Narragansett Bay, where mussels were transplanted to four sites along an established pollution gradient. A broader review of case studies has been produced by Lack and Widdows (9), who conclude that there does appear to be a good relationship between physiological and biochemical responses and tissue concentrations of contaminants, particularly the aromatic hydrocarbons.

24. Further field studies using mussels are being carried out by WRc, including the two case-study long sea outfalls. At Weymouth, mussels were deployed at 3 sites around the outfall:

      site 1 - 'clean' (1.6 km up-current from the outfall)
      site 2 - above outfall
      site 3 - 1.2 km from the outfall, in the direction of
                the prevailing current

25. Results, shown in Fig. 6 indicate that there was no evidence of physiological stress imposed by proximity to the outfall, rather the SFG was significantly higher near the outfall, probably as a result of the good food supply provided by the sewage. However, despite the obvious benefit of the sewage in terms of SFG, there was evidence of some slight metal stress in these mussels. The degree of stress cannot be quantified at this stage and further investigations are being carried out, particularly metal analysis on the mussel tissue the better to relate cause and effect. Unpublished results of mussel transplants around the Littlehampton outfall (Southern Water) indicate similar effects.

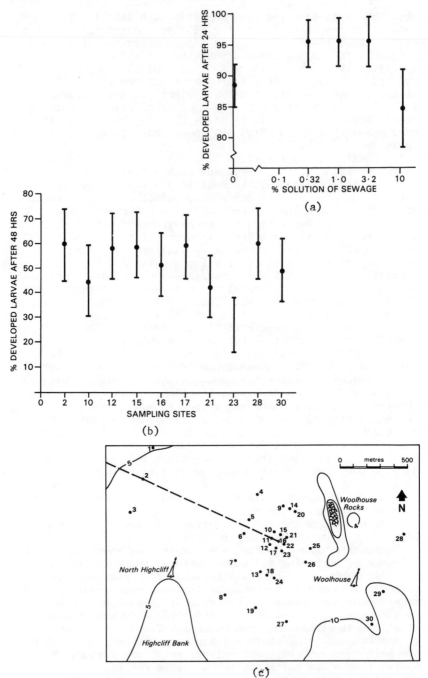

Fig.5. Tenby outfall-bioassays: (a)positions of water sampling;
(b) results of 48 hour mussel larvae bioassay using water
samples collected from around outfall; (c)results of 24 hour
mussel larvae bioassay using sewage dilutions

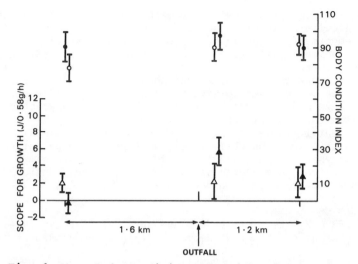

Fig. 6. Mean Body Condition Index (circles) and SFG
(triangles), with 95% Cl in Mytilus edulis exposed
near surface (open symbols) and near bottom (filled
symbols) at three sites in the vicinity of the
Weymouth and Portland sewage outfall

THE WAY FORWARD FOR ENVIRONMENTAL MONITORING AND CONTROL

26. Dispersal of sewage and sewage sludge in the sea is an
efficient and cost-effective means of disposal, provided the
site is well chosen and, in the case of sea outfalls, the
outfall is well designed. The use of computer models allows
the designer to predict the area of impact of various outfall
schemes and choose the most cost-effective one. The
dispersal of sewage sludge can also be modelled. The use of
computer graphics not only aids the engineer, but is a useful
public relations tool because the results are displayed in a
form which is easily understood. Computer models are
currently being used particularly to determine bacterial
impact on bathing beaches and the use of predictive models is
likely to cover the dispersion of a broader range of
contaminants taking into account the partitioning of
contaminants between the dissolved and 'particulate'
fractions of sewage and sludge. Predictive models of the
toxicity of contaminants and the consequences of this for the
long term stability of benthic populations and communities
are also being developed by WRc in collaboration with IMER.

27. Dispersal and die-off of bacteria from sewage and
sludge is well understood and can be predicted. However, the
fate of viruses in the environment is of increasing public
concern and new methods of routine analysis for viruses and
their use on case study sites is of great importance to the
future improvement of public health aspects. The
epidemiology of infections allegedly caused by bathing in
unclean waters will receive greater attention than in the
past.

28. The new approaches to assessing the short and long term biological effects of sewage and sludge disposal to sea, ie meaurements at the cellular, tissue, organ, physiological and autecological levels will become more broadly applied. At present there is a lack of numerical standards for biological effects and these must be established without delay. Such standards may be used to give warning of change or to define acceptable and unacceptable limits. Quality standards for specific contaminants in the biota and sediments will supplement the EQSs already promulgated for contaminants in the water column and collectively will provide a yardstick against which the acceptability of any waste disposal operation can be measured allowing soundly based control measures to be taken when necessary.

REFERENCES
1. TYLER J.M. The distribution and survival of rotaviruses in the environment. WRc/WWA Research Project Annual Report 1985-86.
2. COUNCIL OF EUROPEAN COMMUNITIES. Directive concerning the quality of bathing water (76/160/EEC), Official Journal of the European Community, No. L31, 1976
3. COUNCIL OF EUROPEAN COMMUNITIES. Directive concerning the quality required of shellfish waters (79/923/EEC), Official Journal of the European Communities, No. L281, 1979.
4. WALKER, R.W., WUN, L.K., LITSKY, W. and DUTKA, B.U. Coprostanol as an indicator of faecal pollution. Crtical Reviews in Environmental Control, 1982, 12, 91-112.
5. McCALLEY, D.V., COOKE, M. and NICKLESS, G. Coprostanol in Severn Estuary sediments. Bulletin of Environmental Contamination and Toxicology, 1980, 25, 374-381.
6. BROWN, R.C., and WADE, T.L. Sedimentary coprostanol and hydrocarbon distribution adjacent to a sewage outfall 1984, 18, 621-632.
7. BAYNE B.L. (ed). Cellular toxicology and marine pollution. Marine Pollution Bulletin, 1985, 16, 4, 127-169.
8. WIDDOWS J., PHELPS D.K. and GALLOWAY W. Measurement of the physiological condition of mussels transplanted along a pollution gradient in Narragansett Bay. Marine Environmental Research, 1981, 4, 181-194.
9. LACK T.J. and WIDDOWS J. Physiological and cellular responses of animals to environmental stress - case studies. In: G. Kullenberg (Ed), The role of the oceans as a waste disposal option, 1986, 647-665.

ACKNOWLEDGEMENTS
   The work reported in this paper on sub-lethal effects on caged mussels was carried out under contract to the Department of the Environment whose permission to publish is gratefully acknowledged.
   We are also obliged to Thames Water for allowing us to use information from the Thames sludge dispersal ground and to Southern Water for the use of their data on the Hythe sea outfall location.

15. Applications of research data to the design and performance monitoring of long sea outfalls

J. A. CHARLTON, MSc, PhD, MICE, MIWES, Coastal Engineering Consultant (formerly Senior Lecturer, University of Dundee), and P. A. DAVIES, BSc, DipEd, PhD, Senior Lecturer, University of Dundee

SYNOPSIS. The results of a combined laboratory and field programme of investigations into the phenomenon of salt water intrusion in long sea outfalls, and the consequent effects upon their performance, are described and a hydraulic analysis of the purging process is presented for soffit- and invert-connected multi-riser sections. Laboratory data on the form of the head:discharge characteristics of the outfall system and the dependences of effluent plume trajectories and minor loss coefficients upon outlet port geometry are illustrated. Anti-intrusion and intrusion-limiting devices are discussed, and methods developed for outfall performance monitoring are described.

INTRODUCTION
 1. The use of efficient long sea outfalls as a means of utilising the marine treatment potential of our coastal waters is now generally accepted by public health engineers and marine scientists. However, the responsibility of designing these outfalls to function efficiently remains the responsibility of engineers, who will utilise laboratory and field data provided by multidisciplinary teams. The hydraulic performance of an outfall, while apparently relatively simple when it is regarded as carrying a single liquid discharge, is in fact more complex, since in reality it operates as a device containing two liquids (effluent and sea water) of different densities. One consequence of this is the ability of the denser sea water to intrude into the outfall system and generate hydraulic conditions that can seriously reduce the overall efficiency of the outfall.
 2. A realisation of these problems by the research group at the University of Dundee was recognised by the SERC in 1982 through their support and funding of a major research programme there to investigate sea water intrusion in outfalls. The programme involved extensive laboratory tests on model outfalls in a specially designed facility, (ref.1) and field observations on actual outfalls. This paper describes both of these approaches to solving some of the problems associated with sea water intrusion in sea outfalls.

QUALITATIVE OBSERVATIONS

Apparatus
   3. One of the first products of the research programme was
the understanding gained by qualitative flow visualisation of
the complex processes of sea water intrusion and two density
flow behaviour within an outfall. The sea water intrusion
processes were studied initially with the aid of a laboratory
hydraulic model consisting of (i) a large (6 m^3) sea tank
containing salt water, (ii) a long tunnel with a multi-riser
diffuser section connected to the base of the sea tank, and
(iii) a regulated freshwater supply from a constant head tank
to the upstream end of the tunnel. The sea tank and the
various components of the riser and tunnel sections were
fabricated with clear acrylic material to facilitate good flow
visualisation and the salinity of the water in the sea tank
could be controlled automatically by means of a brine injection
system linked to a salinity probe immersed in the tank. The
freshwater flow through the tunnel was measured with flow
meters, and the intrusion processes were visualised by adding
red Rhodamine B dye at specified locations in the system. The
tunnel:riser section was constructed in modular form to
(i) enable both soffit and invert connections to be studied
and (ii) allow the intrusion and dilution characteristics of
diffuser ports of different geometries to be assessed.

Turbulent mixing
   4. Early tests demonstrated the reluctance of salt and
freshwater layers to overcome their gravitational stability
and mix, unless considerable turbulence is generated at the
interface between them. In the stratified system found in an
intruded outfall tunnel, the necessary turbulence has to be
generated by shear between the upper and lower liquids. This
creates interfacial Kelvin Helmholtz billows which, at higher
velocities break and form entrainment eddies. In plunging
flows, as when an intruding riser feeds into a diffuser, there
is a strong tendency for the denser sea water to fall back
and form a stable stratified layer.
   5. The models demonstrated clearly that in a multiriser
tunnelled outfall system, one of the best mixing and entrainment situations is when fresh water discharging through an
invert-connected riser mixes with stratified static sea water
at entrances and bends. Buoyant turbulent mixing also occurs
freely in an upward discharging riser provide there is a
supply of salt water at its base. (Fig. 1)

Primary and secondary intrusion
   6. Primary intrusion in an outfall is very similar to the
salt wedge effect which occurs in estuaries and channels where
fresh water of mean velocity V and density ρ meets more dense
sea water of density $(\rho+\Delta\rho)$. The appearance of the salt wedge
which develops in the port of diameter D will vary with the
port geometry (Fig.2a), but the discharge at which this flow

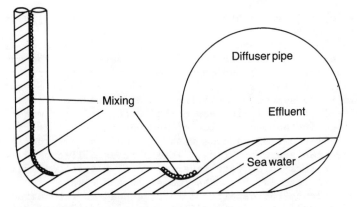

Fig.1 Mixing zones in a purging invert-connected outfall

is first initiated corresponds to the criterion that the densimetric Froude number, F_Δ of the port is less than unity. (Here F_Δ is defined as $F_\Delta = V/(g(\Delta\rho)D/\rho)^{\frac{1}{2}}$, where g is the acceleration due to gravity.)

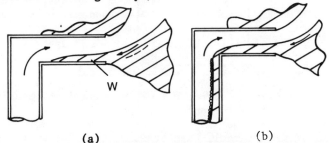

(a)   (b)

Fig. 2 The development of primary intrusion
(a) and the initial of secondary intrusion
(b) in a diffuser outlet port. Salt water wedge,W.

Tests showed the discharge configuration depicted by a solid line in Fig.2a will correspond to an orifice discharge densimetric Froude number somewhat less than unity, while the dashed line in Fig. 2a indicates the edge of the discharging plume for $F_\Delta >1$ cases.

7. Provided that this wedge remains confined to the horizontal section of the discharge pipe, the intrusion process is completely reversible. The position of the head of the wedge within the port is governed by the liquid density difference, the pipe diameter and roughness, and the discharge rate, with inward and outward excursions of the wedge being favoured by low and high discharges respectively.

8. If the discharge reduces to a value where the wedge flow spills into a vertical riser section of the system, secondary intrusion occurs. (see Fig. 2b) The consequences of secondary intrusion are rather more far reaching than with primary intrusion since if one riser in a multi riser system, (Fig. 3)

(in which all risers are initially discharging equally),
becomes subject to secondary intrusion, the flow in that riser,
due to the greater density of the intruded fluid, becomes out
of balance with its neighbours. As a consequence the upward
flow in that riser slows down, the intrusion accelerates, and
eventually dominates, so that the riser goes into reverse flow.
The intruded sea water will then stratify in the diffuser pipe
or tunnel, gradually increasing in depth until sea water is
entrained in the upward flow of another riser. This riser in
turn becomes out of balance with the rest and eventually go
into reverse sea water flow. This process will continue, with
further risers going into reverse, until a state of hydraulic
balance is reached, when the losses in the upward discharging
risers balance the density difference head in the intruded
risers. One or two risers will continue to discharge mixed
fresh and sea water which is entrained from the stratified
layer, and thus satisfy continuity of the sea water flows.

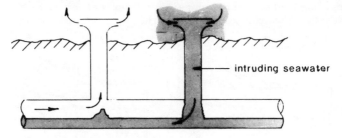

intruding seawater

Fig. 3   The development of secondary
sea water intrusion

The purging of sea water from an outfall

9. If only primary intrusion has taken place (such as would
occur in a plain sea bed pipe outfall with ports in the pipe
wall) then purging is a reverse process of primary intrusion.
If secondary intrusion has occurred with vertical transfer of
saline water, then the purging process is not exactly
reversible. A higher discharge than that at which intrusion
takes place will be needed to affect purging. Moreover, if
purging is not completed in one flow cycle a decrease in
discharge will allow intrusion to re-establish itself.

Outfalls with soffit- or invert-connected risers

10. In an outfall which is discharging normally there is
little to choose between these two configurations, and
intrusion, if it occurs is a similar process in each case.
However, it is in the sea water purging process that the
invert-connected outfall behaves advantageously.

11. A soffit-connected outfall that is discharging in a
balanced intruded condition is depicted in figure 4. A small
increase in discharge will alter the hydraulic balance so that
more salt water is entrained in the upward discharging risers,
the salt wedge will travel downstream, and fewer risers will

be intruded. The speed of the process is controlled by the
rate of increase of discharge and by the rate of sea water
entrainment, which in this configuration is slow.

Fig. 4

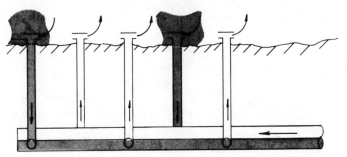

Fig. 5

Figs. 4 and 5 showing intruded outfalls for
soffit- and invert-connections respectively.

From a similar starting point in an invert-connected outfall
(Fig. 5) there is substantially less sea water in the diffuser
initially, and entrainment is accelerated by interfacial
turbulence at the riser entry and bend (Fig. 1). Additionally,
the number and operational sequence of upward flowing risers
can be controlled at the design stage by varying the level of
the riser offtakes.

QUANTITATIVE RESULTS

Hydraulic properties
12. When (i) secondary intrusion occurs and a formerly-
active riser ceases to discharge, or (ii) purging of an
already intruded outfall is effected, concomitant changes
occur in the driving head:discharge relationship for the
outfall system. This is simply because the driving head is a
function of the new velocity head established by either
intrusion or purging, and the friction and minor losses in the
system. The magnitudes of each of these parameters depend
upon the effluent discharge and the number of risers in
operation. The head:discharge curves for 1,2,3 and 4 risers

247

respectively discharging "effluent" in the model are shown in Fig. 6. In the determination of these characteristics, driving heads were measured with sensitive differential pressure transducers (ref. 2).

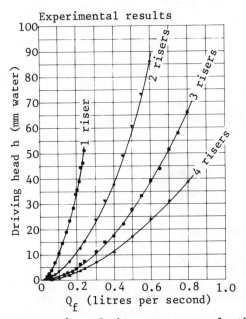

Experimental results

Fig. 6. System characteristics for various numbers of discharging risers

13. Inspection of the curves reveals the discontinuous changes which must occur between them as successive risers are purged or intruded: in particular, for a constant purging flow, the head will follow initially the one riser characteristic until the second riser is brought into operation. At this stage, the head will be forced to transfer laterally to the 2 riser characteristic, and so on. In order to reach the fully purged condition, therefore, the head:discharge relationship will have a step-like form. An illustration of this behaviour is shown on Fig. 7. Note that though the discussion has been posed so far in terms of purging flows, the step-like behaviour is also observed (in reverse) when successive risers succumb sequentially to secondary intrusion. The verification in the model of the form of the predicted head-discharge signature for purging or intruded cases offers encouragement that this property of the system can be used for detecting intrusion in prototype conditions.

Diffuser port design

14. For many design purposes the familiar trajectory diagrams of Abraham and Brooks (ref. 3) may be used to predict plume widths and dilutions. However, little information is available for the regions of the diagram relating to shallow receiving water depths and low densimetric Froude number flows respectively. In particular, it is not known whether the

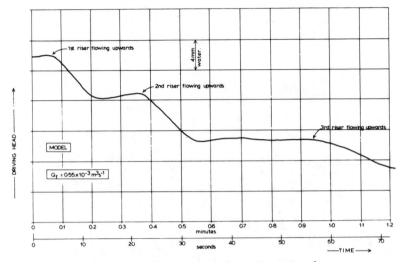

Fig. 7 Model results of purging flow showing step-
like behaviour of driving head as individual risers
are purged sequentially.

trajectories for this region of parameter space are sensitive
to outfall port geometry, as might be expected if momentum
effects dominate the near-field dynamics of the discharging
effluent plume. Trajectories have been measured (ref. 4) for
a number of diffuser port geometries (see Fig. 10a) from
photographic images such as Fig. 10b.

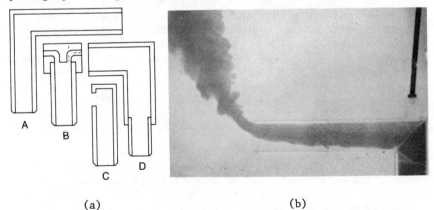

(a)   (b)

Fig. 10 (a) Outfall port geometries used in the study, and
(b) photograph showing discharge from outlet port type A

The resulting trajectory diagram is given in Fig. 11, from
where it can be seen that for quiescent receiving water the
port geometry affects in a significant way the trajectory of
the plume (and hence its surface dilution) for low values of
F_Δ and shallow water.

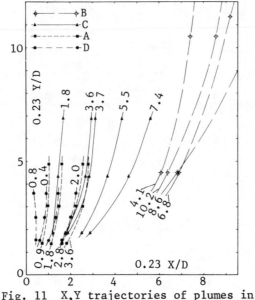

Fig. 11 X,Y trajectories of plumes in
water depth, D for port units A,B,C,D.
Numbers shown are values of F_Δ.

15. In the formulation of numerical models of long sea
outfalls, it is necessary to specify not only friction factors
for the tunnel and risers but also the minor loss coefficients
for bends and outlet port units. The values of these minor
loss coefficients may be estimated from Miller (ref. 5), but
it is not clear a priori that such estimates are valid for
cases in which no long uninterrupted sections are available
before and/or after the component under consideration.
Experiments (ref. 4) conducted with several model outlet ports
have established that there are no significant differences
between the measured coefficients and those calculated by
extrapolation from (ref. 5). Further, when outlet port
lengths are sufficiently short for frictional effects to be
negligible, decreasing length causes increasing minor head
loss.

CONTROL AND PREVENTION OF INTRUSION
16. Sea water intrusion in an outfall is acceptable in
many cases, provided that it can be controlled. Typically,
in a pumped outfall intrusion will occur during the no flow
phases between pumping cycles. Provided that all the sea
water can be purged during the pumping cycles, and flushing
flows achieved, problems due to intrusion are unlikely to
arise. In gravity-discharging outfalls equilibrium
conditions may be reached at low flows when some risers may
be intruded. Once again, provided that a regular daily
purging and flushing flow can be achieved, there are unlikely
to be problems.

17. The ability of an outfall to purge intruded sea water is a function of the capacity of the outfall's discharge regime to maintain the necessary purging flow for a sufficient time for all sea water to be expelled. To achieve this the purging flow must be sufficient to generate losses in the penultimate riser equal to the fresh-sea water differential head developed by the last riser. i.e. The purging flow is, other things being equal, a function of the riser height (ref 6) It follows that tunnelled outfalls, which generally have long risers, are more susceptible to intrusion problems than piped outfalls with short risers. In the case of tunnelled outfalls, the intrusion problem may be aggravated by the tunnel being constructed uphill towards the diffuser. This configuration can allow intruded sea water to flow downhill towards the landward shaft, the additional sea water storage capacity of the tunnel then making total purging far more difficult to achieve.

THE CONTROL OF INTRUSION

The venturi hydraulic block

18. The susceptibility of tunnelled outfalls to intrusion problems can be reduced significantly by including a venturi constriction in the tunnel just upstream of the diffuser. The venturi throat, by locally accelerating the flow and increasing the critical densimetric Froude number, can effectively block sea water from migrating down the tunnel, and thus reduce the volume of stored sea water. This action will enable a much shorter period of purging flow to restore the outfall to outlet port intrusion control conditions. After zero flow conditions when the tunnel will fill with sea water, the venturi can act in an intermittent fashion, enabling successive short duration purging flows to eventually achieve total purging. (Fig. 13)(ref. 7)

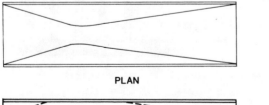

PLAN

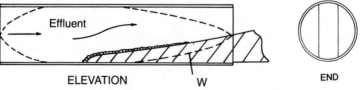

ELEVATION   W   END

Effluent

Fig. 13 Details of the vertical-sided venturi block

Diffuser ports

19. Whereas the tunnel venturi may be considered as a back-stop control, the same principle can be utilised at the discharge ports. By shaping the port as a diffuser (Fig. 14) primary intrusion control is moved to the throat, which may have a minimum diameter while maintaining discharge though a maximum sized port. This configuration will maintain maximum rising plume dilutions over a wide range of flows and reduce overall losses. (ref. 7)

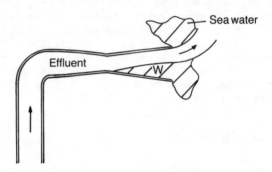

Fig. 14 Illustration of the venturi principle applied to an outfall discharge port.

The cranked tunnel

20. This configuration may be considered as an alternative to the tunnel venturi. Intruded sea water will be stopped at the interface as shown in figure 15. The vertical offset should be greater than any wave action differential, or tidal range if a balancing flow cannot be guaranteed.

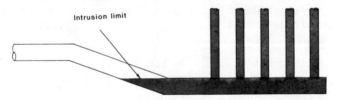

Fig. 15 A cranked tunnel diffuser serving as an intrusion block.

INTRUSION PREVENTION

21. Under certain circumstances total intrusion prevention may be required.

Valves

22. The use of moving valves, although initially the obvious answer, is not generally acceptable with operations departments mainly for reliability and maintenance reasons. However the Taylor-Dunlop flexible duck bill valve (Fig. 16) has been

successfully used on the Weymouth outfall and is expected to
be used on a number of other outfalls currently being
constructed or designed.

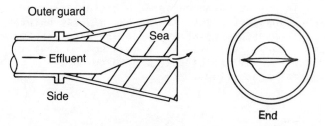

Fig. 16   The Taylor-Dunlop flexible duck bill valve

Siphon exits

23.   The use of an inverted siphon exit (Fig. 17) will act
in much the same way as the cranked tunnel and the overlap
will have to be similarly sized.   It will work well for single
outlet outfalls, but multiple siphon outlets must all work in
horizontal harmony.   Consequently they may fail due to unequal
operation arising from gas accumulations or other factors.

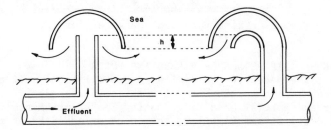

Fig. 17   Two versions of anti-intrusion siphon
outlets

Downward discharging ports

24.   Although not always a practical diffuser design, the
simplest form of intrusion prevention is to use downward
facing ports in a horizontal above bed diffuser as depicted in
figure 18.

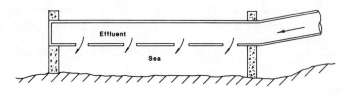

Fig. 18   The invert port non-intruding outfall

THE USE OF INTRUSION IN OUTFALLS

25. It is possible, provided that sufficient energy is available, to utilise controlled intrusion as a prediluting agent within an outfall. Figure 19 shows how this has been achieved with the aid of model tests in one small two-port outfall. Intrusion is encouraged by having a larger port in the invert connected riser, mixing takes place between the two risers and a pre-diluted effluent is discharged through the end riser. The port sizes can be designed so that intrusion takes place for low discharges, say up to DWF, and both ports discharge at higher flows.

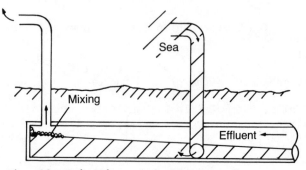

Fig. 19 Using intruded sea water for pre-dilution in a 2-port outfall

THE COMPOSITE DEISGN OF RISERS AND PORTS

26. The necessity to design an outfall to be self purging introduces the possibility of considering the design of the riser and port(s) as a single unit. The requirement is for this unit, at purging flow, to generate losses greater than the riser-fresh water differential head. The losses in the unit (Fig. 20) between 1 and 2 are:-

$$H_f = V_j^2/2g - V_p^2/2g + \text{total conduit losses, } H_\ell \qquad (1)$$

and the purging criterion is $H_f > L(\Delta\rho)/\rho$ \qquad (2)

With H_f fixed, equation (1) may be balanced by varying V_j or H_ℓ. As there is an advantage to be gained by reducing V_j (in order to improve initial dilution), there is scope for reducing the riser diameter. This, in turn, will transfer primary intrusion control from the port exit to the riser itself or the upper bend. The limiting factor in this approach is the minimum acceptable riser diameter, which is probably in the region of 100 to 150 mm for well-disintegrated and screened effluent.

MONITORING

27. A number of methods are available for monitoring the number of outfall risers in operation. Direct monitoring can be

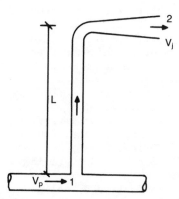

Fig. 20 A riser-port unit

achieved by diver surveys,and indirect evidence for riser malfunction can be obtained by adding Rhodamine dye to the effluent at the point of entry to the outfall shaft. The discharged effluent can then be tracked in the receiving water by means of fluorometers carried on board vessels traversing the line of the outfall. Synoptic information on the number of active risers can be obtained from remote sensing data (ref.8) gathered by multispectral scanner instruments overflying the outfall sites. Fig. 21 shows a processed image of the surface sewage slick above a tunnelled multi-riser outfall, and the line of risers can be discerned clearly. The authors are currently developing a further indirect monitoring instrument, which operates on the principles outlined in para 12. The driving head is obtained from a pair of pressure transducers installed in the outfall and the discharge is recorded by conventional means. The head:discharge relationship is thereby monitored continuously to detect the "steps" in the signature of the outfall's characteristic which are indicative of transitions between n and (n±1) risers in operation. Preliminary tests of the instrument on the Lossiemouth outfall are encouraging.

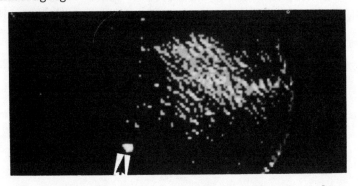

Fig. 21. Processed remotely-sensed image of surface sewage slick above long outfall. Line of active risers indicated

255

ACKNOWLEDGEMENTS
The authors are grateful for the support of the SERC through
the award of grants GR/CO 5236 and GR/CO 5274, and they
acknowledge the technical assistance of Mr M A Breen. The
contributions of Research Assistants Mr G H M Bethune and
Miss L M Macdonald, and Mr S G Dickson, to the success of the
investigations is also gratefully acknowledged.

REFERENCES
1. CHARLTON, J.A. Hydraulic modelling of saline intrusion
    into sea outfalls. Proceedings of International Conference
    on Hydraulic Modelling of Civil Engineering Structures,
    University of Warwick, U.K., 349-356, B.H.R.A. Fluid
    Engineering, Cranfield, U.K., 1982.
2. BETHUNE, G.H.M. A laboratory investigation of saline
    intrusion,with particular reference to its effect on the
    performance of long sea outfalls. M.Sc. dissertation,
    University of Dundee, U.K., 1986.
3. CHARLTON, J.A. Sea outfalls. Developments in Hydraulic
    Engineering - 3,79-128, (ed. P.Novak), Elsevier, London,
    1985.
4. DICKSON, S.G. An investigation of the behaviour of waste
    water flows discharged into quiescent, homogeneous
    receiving waters of greater density, B.Sc. Honours year
    project dissertation, Department of Civil Engineering,
    University of Dundee, U.K., 1986.
5. MILLER, D.S. Internal Flow Systems, BHRA Fluid Engineering
    Series, 5, BHRA Fluid Engineering, Cranfield, U.K., 1970.
6. CHARLTON, J.A., DAVIES, P.A. & BETHUNE, G.H.M. Sea water
    intrusion and purging in multi-port sea outfalls.
    Proceedings of the Institution of Civil Engineers, Part 2,
    1987, vol. 83, Mar., 263-274.
7. CHARLTON, J.A. The venturi as a saline intrusion control
    for sea outfalls. Proceedings of the Institution of Civil
    Engineers, Part 2, 1985, Vol. 79, December, 694-704.
8. DAVIES, P.A., CHARLTON, J.A., BETHUNE, G.H.M. &
    McDONALD, L.M. The application of remote sensing techniques
    to the monitoring of a sea outfall system. International
    Journal of Remote Sensing, 1985, Vol. 6, 967-973.

16. Design and construction techniques for the future

J. M. REYNOLDS, MA, MICE, FIPHE, FIWES, Managing Director, and D. A. WILLIS, BSc, MICE, Chief Engineer, Land and Marine Engineering Ltd, Bromborough

SYNOPSIS.   This Paper deals with engineering aspects of outfall design and construction, describing techniques currently in use, and those which may be further developed into the future.

INTRODUCTION

1. The first long sea outfall in Europe was constructed in Britain, in the Bristol Channel in 1959 with a single discharge at the end of the pipeline.   Five years later a multiple outlet diffuser system was developed for The Hague outfall, heralding the start of what has now become accepted practice for the dispersion and treatment of sewage and trade effluents into the sea.   Since those days many long sea outfalls have been constructed (Reference 1) incorporating knowledge gained over the years together with technical developments and improvements in diffuser arrangements, to achieve a better quality of discharge and selection of the discharge location.

2. Long sea outfalls have been constructed with diameters up to 4 m, and lengths extending offshore to as far as 10 km and water depths approaching 100 m.   While these dimensions are undoubtedly excessive for the great majority of outfalls, it is important to appreciate the inter-relationship between the processes of design and of installation for both small and large outfalls, and that engineers recognise the necessity of thoroughly investigating the marine environment of tides, waves, currents, and often mobile seabeds, in which outfalls will be constructed.   Failure to do so may well lead to an inappropriate choice of material or construction method which could result in unnecessarily high capital and future maintenance costs.   Experience with the construction of all types of submarine pipelines has shown that the foreshore

and tidal zone, to at least 1 km offshore, present unique difficulties which require particular care and skills to overcome.

3. The task confronting the outfall engineer is to design a pipeline between two predetermined positions :

- The discharge location, which will have been selected after water movement investigations and dispersion considerations have been considered such that environmental quality objectives are achieved.

- The headworks which would normally be situated at or near to the outfall pipeline landfall.

It is important at the conceptual stage to study the feasibility and associated budget construction costs of alternative routes that may be considered between discharge and landfall sites so that the selection of the final route avoids unnecessary expense and construction risks.

4. The design and construction processes are considered under four main headings :

- Surveys.
- Design.
- Materials.
- Construction.

These four subjects are all closely related and their order does not necessarily indicate their importance or the procedure that is adopted for any specific outfall.

SURVEYS

5. Having selected the general location of the discharge and landfall areas referred to in paragraph 3, it is necessary to provide environmental survey data and site investigation information. This would be used to evaluate the following :

- The optimum route and outfall profile.

- The outfall security in terms of stability during construction and when in use.

- The outfall material with due regard to the temporary and permanent stresses, alongwith the corrosive or erosive nature of the effluent.

• Construction techniques that can be used to install the outfall, with reference to the specific materials and design.

6. The data and information to be collected would consist of the following (Reference 2) :

Information from Existing Records :

• Coastal erosion.
• Past soundings, seabed levels.

Data Collection :

• Winds.
• Waves.
• Tides and currents.

Site Surveys :

• Bathymetry.
• Geological investigation.
• Tidal streams and current measurements.
• Land survey.

7. A brief outline of a typical survey would be as follows :

Bathymetry :

Soundings would be taken along the full length of the proposed route and over a suitable area either side of the centreline (say 20 m to 30 m) to enable a contour plan to be produced with seabed levels reduced to a suitable datum such as Ordnance Datum Newlyn. The contours should be close enough to allow a longitudinal profile to be drawn along the proposed centreline.

Geological Investigation :

The object of the geological investigation would be to identify the nature of the seabed and its overall geology. This can be achieved by using geophysical techniques, i.e. seismic reflection/refraction and side scan sonar equipment, and by physical sampling with vibrocores or boreholes. The most economic combination is to carry out a seismic and side scan sonar survey with sampling at specific points chosen from the interpretation of the seismic survey. The investigation should cover the same plan area as the bathymetry and should identify the geology to an overall depth below the seabed that will accommodate the full trench requirements (say 5 to 10 m).

Tidal Streams and Current Measurements :

These are usually recorded at two or three locations along the outfall route over Spring and Neap tidal cycles, near to the seabed, at mid-depth and near to the water surface. The measurements are used with wave predictions to enable hydrodynamic loads to be calculated. In certain circumstances, often associated with the water movement studies, it may be appropriate to deploy tethered continuous recording current metres to record currents together with the associated wave effect. The advantage being that the equipment can record over long periods encompassing storm conditions, and thus giving a more accurate statistical assessment of the overall environmental regime. Continuous recording metres are expensive to deploy and maintain, and may only become economically attractive if combined with the needs of the water movement studies.

Land Survey :

The importance of surveying potential construction areas and the landfall site is self-evident, and should be planned to overlap with the marine-based surveys mentioned above, this will ensure that a complete profile of the proposed outfall route can be drawn.

8. The area over which survey data is collected should be sufficient to allow the proposed outfall route to be adjusted so that obstacles or features on the seabed which may affect the construction of a pipeline can be avoided.

9. A second and important further use for the survey data will be by Contractors during the tender period. It is often expedient to produce a separate factual survey report for this purpose.

DESIGN
10. Computers now play an important part in facilitating the design of submarine pipelines and there are a number of sophisticated software systems available to engineers which have been developed to cater for the needs of the offshore oil and gas industry. The engineering philosophy and methods, however, have not changed dramatically and are based on relatively standard theories.

The principle features to (Reference 3) be considered are :
- Stability assessment.
- Security requirements.
- Stress analysis.

The output of the design should be aimed at providing the following information :
- Minimum weight of the outfall to achieve stability against hydrodynamic forces that will be imposed during construction or during the subsequent life of the outfall.

- Outfall profile and minimum burial requirements.

- Minimum structural requirements to enable temporary and permanent stresses to be catered for.

11. Stability Assessment :

Submarine pipelines are subjected to hydrodynamic force due to waves and currents :

- During construction while floating or pulling and before burial or trenching can be achieved.

- When in use, particularily in the event that the seabed erodes or scours to uncover the pipeline.

Long term stability should always be carefully considered if the outfall is to be left on the seabed and not buried or trenched.

The distinction between burial and trenching is that burial includes backfill, whereas trenching defines the situation where the outfall is lowered or placed below the existing seabed level and left uncovered.

The force imposed on the outfall is calculated using momentum equations whereby the water particle velocity or acceleration, due to the combined effect of waves and tidal currents is presented in the following equations :

$$\text{Force} = (\text{Coefficient of drag or lift}) w \, \frac{v^2}{2g} \, d$$

$$\text{and} = (\text{Coefficient of intertia}) \frac{w}{g} \, \frac{d^2}{4} \, a \tag{1}$$

Where
    w = Density of the fluid mass
    v = Water particle velocity
    d = Overall diameter of the outfall
    a = Water particle acceleration

The force is represented by three components :

- Uplift F_u
- Drag F_d
- Inertia F_i

and these are calculated by applying coeffients of uplift, drag and inertia to the appropriate equation shown above. There are differing opinions about the value of the coefficients which, over the years, have been extensively researched, in some cases with physical models.

The authors' prefer to use the following values which originate from research and model studies carried out by Hydraulics Research Limited, Wallingford. (Reference 4).

- For currents : $C_u = 0.5$
  $C_d = 0.9$

- For waves   : $C_u = 1.25$
  $C_d = 1.0$
  $C_i = 3.29$

The water particle velocity is derived by adding vectorially the tidal current velocity and the wave velocity related to a short-term wave return period such as 1:1 year, to cover the construction phase, and 1:100 year return period for long-term stability. Significant wave heights are used throughout with the tidal current velocities measured during the surveys referred to above. The wave height and its associated period are equated to velocity using first order linear wave theory.

The overall force on the pipeline is calculated by adding vectorially the three components of force as follows :

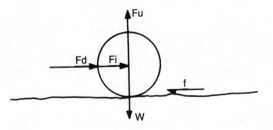

$$\frac{F_d + |F_i|}{f} = W - F_u \qquad (2)$$

Fig. 1

Where

f = Friction factor (varies from 0.75 to 1.5 depending on soil conditions.)

W = Pipe weight underwater.

It should be noted that wave induced forces will oscillate from side to side as the wave passes over the outfall, and will vary in strength sinusoidally according to the wave phase angle.

12. Security Requirements :

The security of an outfall will be affected by the following factors :

- Stability of the surrounding seabed and its resistance to erosion, scour or seasonal changes in level.

- External forces due to fishing or shipping activities.

It is very difficult to quantify the consequences of the above factors and the usual recourse is to choose to bury or trench the pipeline, on the basis that "out-of-sight is out-of-mind". A more rational approach should, however, be made where the stability of the seabed is concerned and in this context the past soundings referred to in paragraph 6 become invaluable. The objective is to estimate the envelope within which the seabed level varies so that the pipeline can be constructed with sufficient cover such that it will not become uncovered during its design life. If this situation can be achieved with a degree of certainty it may be safe to assume that the long term stability requirements referred to in paragraph 11 will not apply.

Other means of providing security are to place stone armouring around the pipeline or to lay mattresses or grout bags over the pipeline. Hydrodynamic forces still apply and it is important to provide sufficient width to cater for the consequences of seabed scour.

13. Stress Analysis :

The scope of the stress analysis will depend upon the construction method or methods that will be appropriate for the site, and the materials chosen for the outfall pipe. The following conditions should be investigated :

During construction :   (Temporary stresses)

- External pressure.
- Handling and fabrication stresses.
- Spanning and curvature.
- Pulling/installation loads.
- Pressure tests.

After construction :   (Permanent stresses)

- External pressure.
- Curvature.
- Backfill loads.
- Internal pressure.

External pressure becomes significant for large diameter pipelines or pipelines laid in deep water, when constructed empty or if it is anticipated that they will be empty during their design life. It can under extreme conditions generate a collapse situation and the "thin wall" pressure vessel elastic stress theory, developed by S. Timoshenko, (Reference 5), will apply for steel pipe, and has been developed to allow for tolerances in out-of-roundness.

Equations are as follows :

$$P_e^2 - [2s(t/D) + (1 + 0.3(D/t))P]P_e + 2s(t/D)P = 0 \qquad (3)$$

Where

P_e = Maximum external pressure for pipe with an initial excentricity of 1%

s = Specified minimum yield stress

P = Critical value of collapse pressure for perfectly circular pipes

$$= \frac{2E}{(1 - Y^2)} \frac{t^3}{D} \qquad (4)$$

E = Modulus of elasticity
t = Pipe wall thickness
D = Mean pipe diameter
Y = Poissons ratio

Hence to avoid collapse, P_e has to be less than or equal to the external pressure generated by the water column at the maximum water depth along the outfall route.

Internal pressures, P_i, due to pumping or gravitional head and the pressures generated during hydrostatic tests, induce longitudinal and radial stresses in the pipe, both of which are calculated using thin wall pressure theory where :

- radial stress $= \dfrac{P_i D}{2t}$ \hfill (5)

- longitudinal stress $= \dfrac{P_i D}{4t}$ \hfill (6)

Spanning, curvature and installation loads generally result in longitudinal stresses and standard linear elastic bending theory will apply for steel pipes. Backfill loads are not normally significant unless large diameter pipes (in excess of 914 mm diameter), are being used, or to accommodate curvature limitations which may cause deep excavations to be necessary in the landfall area. Backfill loads are calculated in the same way as for land pipelines and the resulting radial stress will contribute to the overall structural requirements.

Where the above radial and longitudinal stresses are acting simultaneously they can be combined (Reference 6) using the "von Mises" equivalent stress hypothesis with the following equation :

$$f_c^{\;2} = (f_x^{\;2} + f_y^{\;2} - f_x f_y + 3\,T_{xy}) \hfill (7)$$

Where

$\begin{aligned}
f_c \;&=\; \text{Equivalent stress} \\
f_x \;&=\; \text{Total longitudinal stresses} \\
f_y \;&=\; \text{Total radial stress} \\
T_{xy} \;&=\; \text{Tangential shear stress} \\
&=\; \text{Zero in plane of peak longitudinal stresses}
\end{aligned}$

The equivalent stress is compared with the maximum allowable stress which, in the case of steel would be a percentage of the minimum specified yield stress. There are differing opinions as to what this percentage should be, and a generally accepted criteria for submarine pipelines is :

- 50% for permanent stresses.
- 80% for temporary stresses that occur during construction.

265

The formulae quoted above are for steel or material with similar homogenious properties as steel. Care must be taken when designing with plastic materials to take into account the differing stress/strain relationships, and the stress decay that occurs with age.

Steel and plastic pipe materials will invariably require additional weight for stability against hydrodynamic loads (see paragraph 11) and in the case of steel, this would usually take the form of a continuous reinforced concrete coating. Such a concrete coating is not normally considered as a structural part of the pipeline, and reinforcement is provided to prevent excessive cracking or spalling during handling, and not to enhance the overall strength. The exception to this is for large diameter steel pipelines (above 1067 mm diameter) which are constructed by the Bottom Pull method and will require concrete for stability when the pipeline is empty. As the concrete thickness is likely to be in excess of 150 mm, it is feasible to take it into account when considering external pressures or backfill loads, in order to economise on the thickness of steel.

MATERIALS
14. The materials (Reference 7) that are used for outfall pipelines fall into five main categories :

- Metal - weldable carbon steels.
        - grey or ductile iron.

- Concrete.

- Plastics - Glass Reinforced Plastic.
           - High density Polyethylene.
           - Medium density Polyethylene.
           - Polyvinyl Chloride.
           - Unplasticised Polyvinyl Chloride.

- Flexible armoured pipe.

- Proprietary materials.

The choice of material depends upon a number of factors ranging from corrosive/erosive (Reference 8) properties to construction methods. A further consideration is the means by which individual pipes are jointed to form the overall outfall. Jointing is very important both as regards to quality of the final product, and the feasibility and cost of achieving top quality.

15. Materials currently available are briefly described as follows :

Weldable Carbon Steel :

Steel pipes are widely used for outfall pipelines and are manufactured with either longitudinal or spiral seam welds. The most common standard is the American Petroleum Institute Specification 5L which cover pipelines up to 2000 mm in diameter. British and DIN Standards are also available and all offer a choice of yield stresses which range from mild steel grades to 480 N/mm^2. Pipe is generally manufactured in 6 m or 12 m lengths which can, if required by the construction method, be welded on site to form a continuous steel pipeline. Carbon steels are prone to corrosion both from the effluent internally and the surrounding sea externally. A variety of corrosion coatings can be applied together with cathodic systems, to achieve satisfactory long term protection.

Grey or Ductile Iron :

Grey or ductile iron pipes are manufactured in the U.K. with diameters up to 600 mm and 1600 mm respectively, with lengths up to 8 m. Joints usually take the form of a spigot and socket arrangement with rubber or neoprene sealing rings. Iron pipes have historically usually been associated with short outfalls constructed in the tidal zone and are not generally associated with long sea outfalls. Recently, however, studies (some ongoing) have been made to enable longer lengths to be constructed by means of pulling wires which are attached continuously to the individual lengths of pipe thus enabling long lengths of pipeline to be pulled into position.

Reinforced Concrete :

Concrete pipes are manufactured with diameters up to 4000 mm and usually with spigot and socket "0" ring joints. Sulphate resisting cement is usually preferred and external or internal coatings can be applied to prevent salt water or micro-biological attack. An important advantage with concrete pipes is that they can be made locally to the outfall site in lengths to suit the construction method. Pre-stressed reinforcement or post-tensioned tendons can also be used to increase the radial stiffness or longitudinal strength.

Glass Reinforced Plastic :

G.R.P. pipes are made in a variety of ways either with chopped strand or continuous glass filament reinforcement. The chopped strand type can be combined with a sand/resin filler. G.R.P. can be constructed with spigot and socket joints or as continuous lengths with overlaid or chemically cemented joints. Diameters of up to 4000 mm have been manufactured and specialist resins can be used to achieve internal surfaces that are highly resistant to chemical attack.

High or Medium Density Polyethylene :

H.D.P.E. or M.D.P.E. pipe can be made in diameters up to 1600 mm with lengths to suit construction requirements. The manufacturing process is based on an extrusion technique which, although usually carried out under factory conditions, can be satisfactorily performed on site. Lengths of pipe are joined by heating the ends and pushing the two heated surfaces together to form a butt-weld.

Polyvinyl Chloride or Unplasticised Polyvinyl Chloride :

P.V.C. or u.P.V.C. pipes are usually associated with outfalls where high chemical resistance is necessary. Diameters up to 600 mm are manufactured and joints are usually made with adhesives or by spigot and sockets with rubber 'O' rings.

Flexible Armoured Pipe :

Flexible armoured pipe (Reference 9) has been developed from the submarine cable industry where, by leaving out the central core, a pipe of relatively small diameter is produced. The lining can be of u.P.V.C., and a rubber or plastic outer sheath usually protects the spiral wound steel armouring which provides the longitudinal and radial strength. Diameters of up to 250 mm are manufactured and long lengths of pipe (up to to 1 km) can be made without joints and stored on reels ready for transporting to site.

Proprietary Materials :

Development and research into new types of materials, or variations on the use of the above materials, will continue and has in the past led to manufacturing processes such as corrugated H.D.P.E. and Dunlopipe. Corrugated H.D.P.E. is of particular use with large diameters whereby the corrugations, which run radially around the pipe, add greatly to the internal and external pressure rating, thus making substantial savings in wall thickness. Dunlopipe, now sold as Britpipe, was produced to provide a corrosive resistant material with the strength of steel, and the result is a laminate of alternate layers of steel and resin.

CONSTRUCTION TECHNIQUES
16. Methods used for the construction of long sea outfalls can be grouped under the following headings :

- Bottom pull.
- Float and lower (or flood).
- Laybarge/Reel barge.
- Immersed tube.
- Pipe-by-pipe.

In certain circumstances it may be appropriate to combine two or more of the above methods. Specific techniques such as Horizontal Directional Drilling (Reference 10) or tunnelling (Reference 11) may also be considered to overcome particular problems such as physical features at the landfalls (sand dunes, cliffs, railways, roads, etc.), or with particular reference to tunnelling the geology of the areas.

17. A brief description of the above methods follows and includes reference to materials and any limitations that may apply :

Bottom Pull (Fig. 2.) :

A construction site is formed adjacent to and in line with the route of the outfall and individual pipes are joined to make one or more pipeline strings - their lengths being dependant on the extent of the site. The strings are placed on rollers which are arranged to support the pipeline from the site to the waters edge. Careful weight and buoyancy control of the strings is essential as a small variation can lead to a radical increase in pull loads.

The first string is pulled until the landward end is at the seaward end of the construction site, using a winch or winches mounted on a barge anchored offshore. The next string is then moved across to the rollers and joined to the first string at the tie-in position. The pulling operation is re-commenced and the procedure repeated until all the strings have been joined and pulled into position. Typically a pulling operation will take around one or two weeks, depending on the pipe dimensions and the number of strings. For example, a tie-in joint on a 1000 mm diameter concrete weight coated steel outfall could take up to 8 hours to complete and can add significantly to the overall time for the pulling operation. Pulling capacities of up to 600 Tonnes have been achieved using two winches with multiple purchase pulling systems.

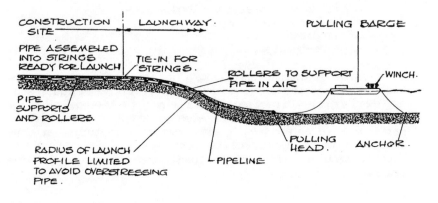

Fig. 2. Bottom pull method

The Bottom Pull method has been extensively used around the British Isles and can be regarded as suitable for exposed coastal locations as well as for sheltered areas. The range of sea conditions during which the pulling operation can be carried out depends on the capabilities of the pulling barge and associated marine craft. In general, significant wave heights of between 2 m and 3 m should be regarded as the limit for safely carrying out the pulling operation.

The traditional material for the Bottom Pull method is weldable carbon steel conforming to API 5L, BS 3601 or equivalent specifications. Steel combines strength in bending and tension to facilitate the installation loads that will be imposed, and has the advantage that it can be continuously welded. It is generally not regarded as good practice to use mechanical joints unless they can be relied upon to transfer the pulling loads without jeopardising the efficiency of the joint. Plastic pipe materials are consequently limited by their low material and joint strengths.

Where steel cannot be adequately protected from the corrosive nature of the effluent, other pipe materials such as u.P.V.C. and G.R.P. with properties that are resistant have been used as a lining to the steel pipeline. The resulting composite pipeline can then be installed by the Bottom Pull method.

Outfalls of up to 3000 mm in diameter have been installed by the Bottom Pull method, with lengths of up to 5 km. Submerged weights for outfalls when empty (pulling condition) usually range between 50 to 200 kg/m. The resulting weight when in service, full of effluent, will considerably improve on-bottom stability, and is, of course, the permanent condition to be considered.

The use of sacrificial wires to take loads generated during pulling operations are currently being investigated with the aim of extending the range of materials that can be used to those with low material and joint strengths. (See paragraph 15 - Grey or Ductile Iron).

A variation of the Bottom Pull method is referred to as the "off-bottom tow" technique, whereby the pipe is fitted with buoys and chains and towed into position. The weight of the chains balances the uplift from the buoys such that the pipe floats one or two metres above the seabed with the chains trailing along the seabed. When in position, the pipe is flooded, the buoys removed and the pipeline consequently settles into position. This technique has not, to the Authors' knowledge, been used for the construction of an outfall but will constitute a method that can be considered for the future if the route is in a sheltered area where a land site long enough to facilitate the full length of the pipeline exists.

Float and Lower (or Flood) (Fig. 3.):

A construction site on land adjacent to sea or a wharf is formed and individual pipes are joined to make one or more pipeline strings. The length of the strings will be dependant on the extent of the site which need not necessarily be near to the location of the outfall. Careful weight and buoyancy control of the strings is essential as small variations can lead to difficulties during the floating and lowering operation.

The completed strings are moved into the water and secured under supporting pontoons spaced to suit the strength and buoyancy requirements of the pipe. For larger diameter

271

outfalls, there may be sufficient inherent buoyancy to enable the strings to float without pontoons. The strings are towed and manoeuvered into position over the location of the outfall ready to be lowered in a controlled fashion into position. The lowering operation is achieved by means of winches mounted on the pontoons, by flooding the string with water or by a combination of the two. If the strings are lowered by flooding, it is recommended that a 'Pig' is used in conjunction with the introduction of water, so that air in the pipe is completely evacuated. If more than one string is required joints are made on the seabed after each subsequent lowering operation using flanges or other mechanical couplings.

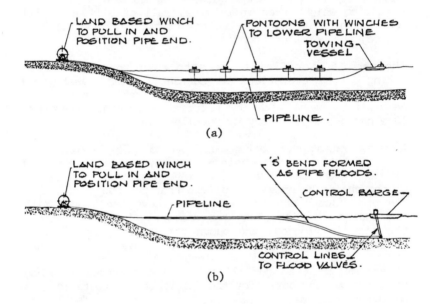

(a)

(b)

Fig. 3(a) Float and lower; (b) float and flood

The float and lower method is usually associated with sheltered shallow coastal locations, estuaries, rivers or lakes, and should be regarded as highly sensitive to weather and sea conditions. The highest risk occurs when the strings are being lowered into position and in general significant wave heights in excess of 0.5 m should be avoided. The risks associated with the floating and lowering operations require a high degree of specialist marine skills and equipment.

The choice of pipe material and jointing method will depend largely on the location of the outfall, and in

general terms, weldable carbon steel conforming to API 5L,
BS 3601 or equivalent specifications, will be preferred
for exposed locations. Plastics such as G.R.P., M.D.P.E.
or H.D.P.E. can however be considered for lakes and
sheltered locations. In all cases the joints will have to
withstand tensile loads as well as bending, and welded or
bolted flange connections are preferred.

Where steel cannot be adequately protected from the
corrosive nature of the effluent, other pipe materials
such as u.P.V.C. with properties that are resistant can be
used as a lining for the steel pipeline. The resulting
composite pipeline can then be installed by the float and
lower method as described above.

There is no limitation on the size of pipes that can be
installed by the float and lower method, diameters of up
to 3000 mm and length up to 1000 m have been successfully
installed.

Laybarge/Reel Barge (Fig. 4. and Fig. 5.) :

Laybarges and Reel barges are usually associated with long
pipelines for deep water offshore projects related with
the oil and gas industry. They are large, sophisticated
vessels that are not normally economically viable for the
relatively short lengths of pipe that are required for
outfalls. In the context of outfall construction the
laybarge or reel barge would be less sophisticated
consisting of a modified flat deck barge which would be
suitable for operating in shallow inshore areas.

A small construction site is required at the outfall
landfall for winches and associated equipment to enable
the outfall pipes to be pulled from the laybarge or reel
barge to the shore. A second area which need not be near
to the location of the outfall will be necessary for pipes
to be assembled, concrete coated and lined in readiness
for shipping to the laybarge. For a reel barge this
second area would need to be suitable for assembling
pipeline into long lengths ready for reeling on to the
barge.

The laybarge or reel barge would be positioned on the line
of the outfall as near to the shore as the operating
conditions and draft of the vessel allow. Pipe would then
be pulled (using the Bottom Pull method) from the barge
through the surf zone to the high water level. The
laybarge or reel barge would then be manoeuvred away from
the shore by means of its own mooring anchors, whilst at
the same time paying-off pipe on to the seabed. The
laybarge and reel barge methods are not commonly

273

associated with outfall construction due to the specialist nature of the barges and their associated plant and equipment. Sea conditions are also important and the laybarge or reel barge must have sufficient anchors to enable them to manoeuvre along the outfall line, whilst at the same time retaining stability to prevent damage to the pipe between the barge and the seabed.

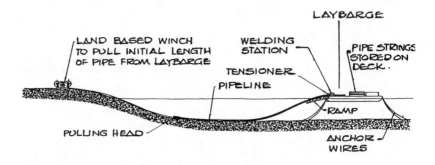

Fig. 4. Laybarge method

Weldable carbon steel and high or medium density polyethylene have been fabricated from laybarges, the latter being more appropriate to sheltered areas where tensile properties are less important. Pipe materials that cannot be continuously welded can also be installed providing there is some temporary means of preventing the joints from opening when under tension between the laybarge and the seabed.

Flexible armoured or high density polyethylene pipes are usually associated with the reel barge method, the latter being more appropriate for sheltered areas where tensile properties are less important.

There is no limit to the length of outfall that can be installed by the laybarge or reel barge method and it would generally be found that they become more economical per metre as the length of the outfall increases. The nature of the reelbarge method dictates that only small diameter pipes can be considered, up to approximately 300 mm diameter, and where larger diameters are required the possibility of laying more than one pipe to achieve the required equivalent discharge quantities can be considered.

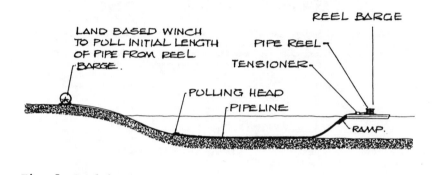

Fig. 5. Reel barge method

Immersed Tube (Fig. 6.) :

The immersed tube method has been developed for the construction of large rectangular or circular culverts which are normally associated with road tunnels and cooling water intakes.

A construction site is required where individual pipes can be manufactured and joined into lengths that are suitable for the type and size of floating plant that will be used. The site does not have to be near to the outfall route, but does require access to the sea and should be in a reasonably sheltered location. The pipe lengths can be floated into the water via a slipway or by making the construction site into a large graving dock which can subsequently be flooded.

Individual pipes are joined into lengths of up to 50 m long. The lengths are fitted with watertight bulkheads so

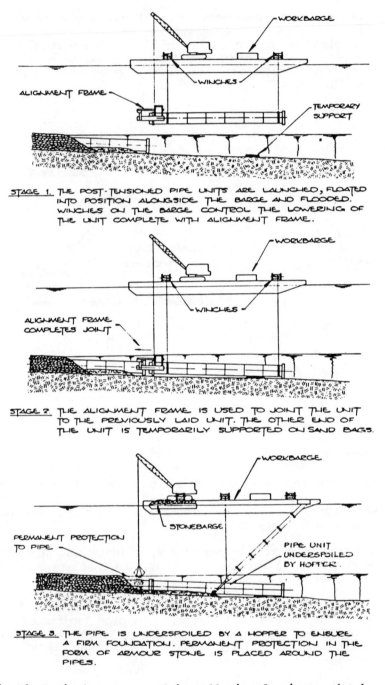

STAGE 1. THE POST-TENSIONED PIPE UNITS ARE LAUNCHED, FLOATED INTO POSITION ALONGSIDE THE BARGE AND FLOODED. WINCHES ON THE BARGE CONTROL THE LOWERING OF THE UNIT COMPLETE WITH ALIGNMENT FRAME.

STAGE 2. THE ALIGNMENT FRAME IS USED TO JOINT THE UNIT TO THE PREVIOUSLY LAID UNIT. THE OTHER END OF THE UNIT IS TEMPORARILY SUPPORTED ON SAND BAGS.

STAGE 3. THE PIPE IS UNDERSPOILED BY A HOPPER TO ENSURE A FIRM FOUNDATION. PERMANENT PROTECTION IN THE FORM OF ARMOUR STONE IS PLACED AROUND THE PIPES.

Fig. 6. Typical sequence of installation for immersed tube method

that they can be floated to the location of the outfall. The lengths are then positioned in a controlled manner by a combination of flooding and lowering under floating plant, into a pre-dredged trench. Weight control is critical in order to ensure that the pipe lengths can be safely handled, placed and joined to the previously positioned pipe length.

Watertight joints are achieved by using spigot and socket 'O' ring joints or specially designed rubber butt seals, and it is usually necessary to use an underwater alignment frame to position and complete the connection between pipe lengths. Backfill is placed around the pipe lengths after they are positioned and foundation material is usually pumped underneath the pipe lengths to minimise settlement.

The immersed tube method is an economical solution for large diameter outfalls, and is normally associated with sheltered coastal locations and estuaries. The sensitive nature of the marine operations required for installing the pipe lengths means that a high degree of specialist marine skills and equipment are necessary. In general significant wave heights of 0.5 m should be considered as a maximum during installation.

Individual pipes are normally made out of reinforced concrete designed so that they can be joined into lengths by post-tensioning techniques. Steel or G.R.P. can also be considered.

There is no limitation on the size of pipes that can be installed by the immersed tube method, although for outfall applications diameters of between 3 m and 5 m would normally be considered feasible.

Pipe-By-Pipe (Fig. 7.) :

Pipes for this method of construction are usually made at a factory, but in some cases can be manufactured at a site specifically created for the purpose near to the outfall location.

The pipes are lowered into their required position on the seabed or into a pre-dredged trench from a crane or gantry mounted on a floating workbarge. Joints usually consist of spigot and sockets with 'O' rings, although in some cases bolted flange connections or proprietary mechanical systems can be employed. For larger diameter pipes an alignment frame with hydraulic rams is usually necessary to enable the pipes to be pulled together and supported while backfill material is placed as a foundation.

The pipe-by-pipe method is generally considered suitable for short lengths of outfall in shallow water sheltered coastal locations, estuaries, rivers and lakes. The rate of progress is limited by the nature of the installation procedure and the overall time to complete the work can often be seriously affected by inclement sea conditions. The range of sea conditions during which the placing operation can be carried out depends on the capabilities of the workbarges and associated marine craft. In general, significant wave heights of 0.5 m should be regarded as the limit for installing the pipes. The nature of the operation also requires a high degree of marine skills including diving. A wide variety of different types of material and joints are feasible for the pipe-by-pipe method, and mechanical joints with no tensile properties can be used. Concrete, G.R.P., high density and medium density polyethylene and other plastic materials are all manufactured with mechancial joints and their selection is usually a function of cost and compatability with the effluent being discharged.

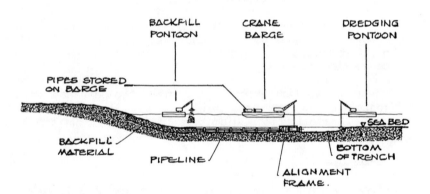

Fig. 7. Pipe by pipe

Burial and Trenching :

Trench excavation can be considered before (pre-lay) or after (post-lay) installation of a submarine pipeline. Conventional dredging plant is invariably associated with

pre-lay excavation, while seabed trenching tools are used
for post-lay trenching.

Trailing suction, cutter section, and bucket dredgers can
generally operate to 30 m water depth, exceptionally to
40 m. They can produce relatively smooth regular trench
profiles in sedimentary beds and some soft rocks. While
the grab dredger can work in greater depths, it tends to
produce irregular profiles.

Post-lay seabed trenching tools, extensively developed for
oil and gas submarine pipelines, can be grouped under the
following techniques and applications.

Sand lifting	–	non cohesive sands
Low pressure jetting	–	sands, silts, and very soft clays
High pressure jetting	–	most sediments up to firm clays
Ploughing	–)	most sediments up to
Rotary Cutting	–)	firm clays. Difficult with non cohesive sands

Usually these seabed trenching tools use the pipeline as a
"guide rail", and many can be deployed, operated, and
recovered without the use of divers; controls and
monitoring being on the attendant vessel. In addition to
their primary soil disturbance mechanism, they all employ
soil excavation or removal devices – with the exception of
the high pressure jetting systems. In the Authors'
opinion the latter should be used with caution – they use
a tremendous amount of power, which may cause over
excavation and pipe spanning, or lowering of the seabed
area rather than just the pipeline.

Generally, a seabed trenching tool, can be used on
pipelines up to 1 m diameter, exceptionally 1.5 m
diameter. They require a relatively smooth external
surface free from substantial protrusions, e.g. flanges or
intermittent concrete rings. They have a trench depth
limitation, at present in the region of 2.5 m, though
continuing development may increase this depth capability.
In comparable situations, and in the smaller range of
outfall sizes, they do offer considerable cost savings
over pre-lay dredging. Also the often considerable risk
of trench siltation is removed.

The siltation of dredged trenches is particularly important where long lengths or the whole of an outfall trench is to be opened up pre-laying, i.e. with the float and lower, pulling, lay or reel barge methods.

However, the selection of a seabed trenching system, or the right type of dredger(s) and attendant feasibility of a trench remaining open, are questions which must be addressed to each project, and can only be done with good knowledge of seabed soil characteristics and the local marine environmental conditions.

A common feature associated with most outfalls is the foreshore and surf zone, where dredgers or trenching equipment cannot operate due to the shallow water. The usual remedy is to use land-based excavation equipment working tidally directly off the foreshore or from a causeway, and it is often expedient to use sheet piled cofferdams to minimise the quantity of excavation and to protect the shoreline or flood defences.

CONCLUSIONS
18. A wide variety of materials are currently available for outfall pipelines and the final selection has to take into account the practicability of construction as well as the corrosive/erosive nature of the effluent to be discharged. The design process, where the selection of the permanent materials is concerned, may well take place before the construction method can be finally decided. This is particularly the case if a design and construct type of contract is envisaged and the possibility of more than one material may well be worth considering in order to afford contractors as much flexibility as possible.

19. The selection of a method (or methods) that will be feasible for a particular outfall location will depend entirely on the circumstances surrounding the proposed route and should be dealt with on a case-by-case basis. The methods described above have proved, by experience, to be feasible and sound for various types of underwater pipelines, and will, in the Authors' opinion, continue to be developed and applied to future outfall construction.

REFERENCES
1. CHARLTON J.A. Contemporary Developments in Outfalls Design. I.P.H.E. Technical and Training Symposium - Construction and Maintenance of Long Sea Outfalls. 1986.

2. WILLIS D.A. Site Investigation and Selection - Engineering Aspects. Proceedings. I.C.E. Conference on Coastal Discharges, 1983.

3. REYNOLDS J.M. Design and Construction of Sea Bed Outfalls. Proceedings. I.C.E. Conference on Coastal Discharges, 1983.

4. Hydraulics Research Station, Wave Forces on Pipelines, H.R.L. Wallingford, 1982.

5. TIMOSHENKO, S. Theory of Elastic Stability. McGraw-Hill Book Company, New York.

6. Det Norske Veritas. Rules for Submarine Pipeline Systems, 1981.

7. CRISP, E.W. and McVIE, A. Materials used in Construction. Proceedings. I.C.E. Conference on Coastal Discharges, 1980.

8. RICHARDSON, L.W. Corrosion Effects and Use of Resistant Materials in Sewerage Systems. Institute of Water Pollution Control; Symposium in Septic Sewerage. Problems and Solutions. Bournemouth, May, 1979.

9. DAVIS, A.L. Flexible Outfall at Aldeburgh. Effluent and Water Treatment Journal, 1980.

10. REYNOLDS, J.M. and SZCZUPAK, J.R. Directional Drilling Experience. International Society for Trenchless Technology No-Dig 87 Conference.

11. MOORE, K.H. Tunnel Outfall Design and Construction. Proceedings. I.C.E. Conference on Coastal Discharges, 1980.

Discussion on Papers 14-16

DR LACK and MRS COOPER, Paper 14
A major area of concern when designing a sea outfall is the
prediction of any adverse effects it may have on the sea bed
and the creatures that live on it. Effects on ecology tend
to occur in benthic rather than pelagic communities. Benthic
invertebrates have relatively sessile lifestyles and cannot
readily escape the effects of pollution by swimming to
another area. The natural environment in which these
organisms live is often hostile and can be completely
disrupted by storms or changes in tidal currents. Because of
this, many benthic organisms tend to be opportunistic in
their lifestyle, i.e. they can invade an area as soon as it
becomes habitable and increase their numbers rapidly. These
naturally large fluctuations in numbers make pollution-
induced changes difficult to detect and virtually impossible
to predict. Changes attributable to sewage outfalls are not
always adverse in terms of total numbers of animals. The
oligochaete worms, in particular, thrive on the high levels
of organic matter associated with sewage. The distribution
of species may change, resulting in a lower diversity, while
the total numbers of individuals may increase.

The se difficulties have led to the recommendation (ref. 1)
that it is better to search for patterns among biological
data and interpret them in terms of environmental data, such
as particle size distribution and organic carbon content of
the sediment. The methods used to reveal patterns and
structure in data are multivariate statistics which aid, but
do not substitute, informed interpretation (ref. 2).

Such statistical methods can assist in demonstrating
adverse effects of an outfall, but they also show patterns
which are due to natural phenomena and not a consequence of
pollution (ref. 3). It is therefore important to carry out
thorough baseline surveys. An example of their value is the
survey carried out by Southern Water at Hythe, where a new
outfall is to be commissioned. The natural variation in
sediment and faunal distribution gives the appearance of a
plume effect originating at the proposed site of the outfall.
If the baseline survey had not been carried out this may well
have been interpreted as an adverse effect of the outfall.

Marine treatment of sewage and sludge. Thomas Telford Ltd, London, 1988

Minimum benthic sampling programmes should include

(a) surveys before and after construction/commission of
     the outfall
(b) bacterial analysis of sediments
(c) particle size analysis of sediments
(d) organic carbon analysis of sediments (e.g. loss on
     ignition)
(e) identification of benthic infauna to species level
(f) multivariate statistical analysis.

Interpretation of ecological data will be severely limited
if there is a lack of determinands and computing facilities.
   Additions to benthic sampling programmes, as finances
permit, should include

(a) viral analysis of sediments
(b) metal analysis of sediments
(c) persistent organics analysis of sediments
(d) sewage specific compounds (e.g. coprostanol) in
     sediments
(e) C:N ratio
(f) metal uptake by plants and animals
(g) sublethal toxicity tests (bioassays)
(h) underwater film/videotape of area immediately around
     outfall before and after commission.

MR P. N. PAUL, John Taylor & Sons
One of the fascinations of marine treatment systems is the
way in which our knowledge of the processes involved, the
hydraulics of the systems, and our ability to predict outfall
performance have evolved over the past 15-20 years. This
evolutionary process is embodied in Papers 14-16. They
recognize that sewage cannot be thrown into the sea, and more
importantly they recognize that the industry must continually
strive to improve its knowledge of the effect of sewage and
sludge discharges on the marine environment. In my view this
is absolutely essential if the UK is to be able to justify to
others that the discharge of sewage and sludge is a satisfactory method of disposal which is not going to create long-
term operational difficulties and environmental problems.
   The exciting work currently being undertaken at WRc falls
broadly into three categories

(a) development of mathematical models to predict
     dispersion of sewage and bacterial die-off
(b) investigations into viral decay in the marine
     environment
(c) environmental impact and assessment of sewage and
     sludge discharges of marine ecosystems.

Mathematical models are valuable tools for predicting the
performance of outfalls and can save much money and time in

undertaking extensive marine investigations. It is, however, extremely important to remember that the models are only predictive tools. So many of the basic parameters in the dispersion model do not have precise values which are known for each location. Actual T_{90} times and horizontal dispersion coefficients are difficult to determine and will vary from one site to another and depend on weather conditions. These and other basic parameters can, however, be readily varied in the model to enable the sensitivity of assumed values to be determined.

Since the models are predictive it is still necessary to carry out field trials using, for example, tracers to prove that the location of discharge proposed from model predictions will be satisfactory. The graphic output from the dispersion model is also extremely useful for public relations purposes, particularly if the results from the proving field trials or from existing discharges can be used to calibrate the dispersion model. The public do not understand mathematical models and tend to distrust them. There is therefore a need to feed field-proving data back into the model so that the final version shows as accurately as possible what will happen when the outfall is commissioned.

The Authors mention the calibration of the hydrodynamic model using float tracks. WRc and my firm are working together on a number of outfall schemes but at one in Blackpool the North West Water Authority has used radar to develop a knowledge of tidal movements and for calibration of the hydrodynamic model. I would be interested to hear the Authors' views on the use of radar for this purpose and whether they consider that this method of determining actual tidal movements has advantages over the more normal float tracking.

I was particularly interested to see the results of the field studies on the effects of the Weymouth outfall and obviously pleased to see that there appears to be little (if any) effect on the marine environment. I am not, however, entirely familiar with the indices used and would be pleased if the Authors could elaborate. In particular perhaps they could indicate what values of 'scope for growth' and body condition index could be considered to be good and what values might show that conditions are unsatisfactory. Viral standards also form part of the EC directory requirements. Could the Authors elaborate on any future work that they will be undertaking?

Paper 15 summarizes the important work that has been carried out at Dundee University in recent years. These studies have been extremely valuable to the industry in helping to understand the hydraulic regimes which can exist in diffuser systems and in helping designers to overcome problems of saline intrusion.

It is important for all designers to appreciate the effects of secondary saline intrusion not only on the hydraulics of the outfall system but also in respect of the sands and sediments that can be drawn into the outfall if intrusion occurs on a large scale. In particular the Authors have been correct to point out that the head and total volume of sewage which may be required to purge outfalls may be greater than that anticipated under normal maximum flow conditions with all risers open. Under certain circumstances, special facilities may need to be incorporated to ensure that the outfall can be purged of sea water.

I am interested in the Authors' proposals for the monitoring of manifold and riser operation as I believe that this will be an important aspect for incorporation in future designs. Could the Authors elaborate on how they believe techniques will develop in the future?

Another interesting possibility maintained in Paper 15 is the concept of incorporating controlled intrusion into an outfall with the result that the discharge from the outfall is already a mixture of sewage and sea water, thus increasing dilution factors. Could the Authors please elaborate and say whether in their view this has scope for further development?

Paper 16 summarizes the factors to be considered in outfall pipe design. However, the basic outfall is not the whole cost of a marine treatment scheme. Pre-treatment of sewage before discharge is essential in most cases. Very often a headworks has to be constructed in environmentally sensitive areas, and as a result screening and grit removal equipment has to be incorporated into an enclosed structure to minimize noise and smell problems. The costs of such facilities are high, often as much as the outfall itself, and an essential feature of a marine treatment scheme.

In view of the high capital cost of outfall schemes (and costs which it is sometimes difficult to phase), I would hope that attention will be paid in the future to ways in which pre-treatment costs could be reduced. In addition, I would hope that the industry might address outfall construction methods and pipeline materials which would enable outfalls to be installed with a lower risk element than is appropriate at the present time. The development of directional boring to enable longer and larger outfalls to be constructed by this method might for example be one possibility. Could the Authors elaborate on the areas of pre-treatment and outfall construction methods which they believe could and should be addressed in an effort to reduce the present relatively high construction costs of outfall schemes?

MR T. D. MacDONALD, Scottish Development Department
It is encouraging to see the enthusiasm that is being devoted to the improvement of conditions in waters off England and Wales; in Scotland such waters have been under control for many years, even pre-dating the Control of Pollution Act, 1974. As a consequence much has already been done.

There is evidence, however, of stress and irreconcilable views. Much sorrowful head-shaking took place about the

unreasonable attitudes of German and other colleagues. We
will have to be more persuasive: if we end up with a ban on
processes we currently find useful, it will be our failure.

Toxic and persistent substances should be kept out of
sewers and outfalls. To state a problem is not to solve it
and I question the extent to which this ideal is achieved.
In this connection I wonder why this Conference has not
considered the control of industrial discharges, many of
which have similar characteristics to the domestic effluents
which do concern us. On storm water overflows, they and
their effects have been mentioned frequently, rather as one
might refer to a distressing, regrettable and unavoidable
social problem. I would certainly advocate a more active
approach to assessing options for living with them.

In conclusion, here is a paradox: if an outfall system
fails to meet its design criteria, those most directly
concerned are probably not too fussed; people living locally
will appreciate that conditions are much improved. However,
if responsible monitoring were to disclose failure, this will
not be lost on the opponents of the UK approach, and scope
for such improvements may well be restricted as a
consequence. The reverse is not true.

MR M. D. McKEMEY, Lewis & Duvivier

My comments refer to Paper 14. The monitoring of
environmental impacts is absolutely vital, as only from the
knowledge derived from monitoring can realistic environmental
assessment of proposed projects be carried out.

Predictions relating to coliform levels at bathing beaches
cover only one aspect of the total environment. The
environment includes ecological, physical, aesthetic, social
and cultural aspects. All these aspects should be addressed
if only to show that they are not significant. By including
the full range of environmental aspects in a sensitivity
analysis to compare different treatment options any
attraction in marine treatment may be more apparent. One
problem when considering, say, the predicted effects of
marine treatment on marine species, is the need to rely on
educated opinion alone. I should be particularly interested
to know whether the Authors consider that impacts on marine
ecology could in some way become amenable to mathematical
modelling in the future.

MR D. V. BUDD, The Kenny-Snook Association

The Authors of Paper 16 refer to submarine pipelines in the
oil and gas industry and I should like to discuss some of the
current engineering practices in this industry. I agree that
the design and construction of submarine pipelines for the
foreshore and tidal zone present unique problems. However, I
suggest that there is considerable common ground in both
offshore and near-shore pipeline engineering which facili-
tates the adoption of some offshore engineering practices for
the benefit of outfall design and construction.

287

The choice of hydrodynamic coefficients for calculating the forces applied to a pipeline under wave and current action can be confusing as the Authors imply because of the large number of data sources available. It is therefore important when choosing the coefficients to consider the method by which stability of the pipeline will be analysed. One method commonly used is given in the Paper; another method at the other end of the scale recognizes that the complete length of the pipeline does not usually experience the peak hydrodynamic loading at the same instance and therefore allows limited deflection under the action of this peak load.

The choice of wave theory to be used in the assessment of stability can have a marked effect on the value of the loading to be applied in the analysis. Use of linear theory with a significant wave as suggested by the Authors is a common approach. However, it may inadvertently ensure that the effect of breaking waves is not taken into account. The large increase in velocity obtained when using solitary wave theory to account for breaking waves can be quite confusing when deciding which velocity to use for the stability assessment. Furthermore, problems can be experienced because of the limited applicability of one wave theory for the large range of water depths experienced with near-shore pipelines.

Because of the nature of the sea bed, spans occur in installed pipelines even though the pipeline route may have been chosen to eliminate this problem. Moreover, installation of a pipeline may generate spans through scouring of the sea bed. It is suggested that spans should not be judged solely on the level of induced longitudinal stress in the pipeline but should be assessed also for vibration, loss of concrete coating, scouring and trawl gear hooking.

Full-scale testing of pipeline spans has been performed in the Severn Estuary and has shown that loss of concrete coating can occur with excessive vibration. Spans may be subjected to vibration through vortex shedding and if not rectified may grow and can lead to fatigue failure.

In addition to hooking of spanning pipelines, trawl gear can impart significant impact to a submarine pipeline lying on the sea bed. With trawl boards being towed at three to four knots with a warp tension of between 5 t and 30 t the trawl board is capable of denting a steel pipeline. These dents may lead to failure by cracking of the internal lining or causing a buckle to be initiated which then propagates along the pipeline.

Use of a concrete coating assists in spreading the effect of the impact. The ability of the concrete coating to withstand repeated trawl board impacts is sometimes tested within the oil industry by setting up a test rig to simulate the trawl board impact and measuring the number of blows required to expose the reinforcement and also the corrosion coating.

This test is simple to perform and can increase the confidence level for a submarine pipeline being installed in a known fishing area. It would not be difficult to extend this test to the protection structures of diffusers.

The levels of stress allowed in the oil and gas industry are generally higher than the values suggested by the Authors. These allowable limits vary depending on the location of the pipeline and the hazard level of the contents. It should be noted that operating conditions are generally more severe in oil and gas pipelines (typically pressures up to 200 bar). A consequence of this is that a better understanding of material properties when subjected to loads was necessary to prevent over-conservative design with consequent cost implications. This increased understanding and more sophisticated stress and strain analysis has resulted in higher values of permissible stress being adopted. I suggest therefore that the choice of an allowable stress level should consider the function of the pipeline and the restraints of the installation method.

The off-bottom tow construction technique has been used successfully for submarine pipelines and is suggested as a technique for marine outfall installation. Where the land site is not long enough to facilitate construction it should be possible to construct the outfall at a suitable location nearby and to tow it into position with final pull-in to the shore.

Laybarges are sophisticated and relatively expensive pieces of equipment but they should not be dismissed as being unsuitable for marine outfall installation. They have been used for shore approaches and have laid relatively short pipelines in water depths of 30 m or less. Use of laybarges can cut down the time required for installation and can be economically viable if already mobilized for other projects. An advantage of this method is that onshore work is considerably reduced. This may be particularly important in coastal towns where limited impact on holiday trade is a major consideration.

The laybarge technique need not be carried out by a large third generation vessel. A flat-bottomed work barge could be adapted to lay pipe in this manner. Costs involved in such work may appear large if related to one project, but when the vessel can be employed on maybe two or three such projects it could prove to be a very beneficial modification.

MR M. KING, Land & Marine Engineering Ltd
A comforting fact that has emerged from Paper 16 is that techniques used for the past thirty years have stood the test of time. No drastic change seems likely - certainly in the immediate future.

Because of the heavy injection of capital involved in a short space of time, the funding of sea outfalls could be and often is an embarrassment to public authorities. The possibility of a total package involving investigation, design, construction, maintenance and funding could be attractive. There might still be problems over repayment within the specified EFLs at a rate which authorities could afford, but in principle the suggestion could have merit. I would like to ask how the Authors view design/construction contracts. Do they welcome this approach and see this service growing in popularity?

Do the Authors see pulled outfalls in ductile iron competing advantageously with traditional welded steel?

MR J. A. WAKEFIELD, Coastal Anti-Pollution League
Do I detect a reference to privatization of the water industry in connection with the long-term finance of a marine disposal system in the form of a turnkey contract?

MRS L. EVISON, University of Newcastle upon Tyne
I would like to endorse the concern in Paper 14 about the problem of enterovirus uptake in filter-feeding shellfish, and the consequent potential for causing outbreaks of viral gastro-enteritis in humans. Studies by one of my research students have shown that although bacteria taken up by mussels grown in contaminated waters can be eliminated within two days in a shellfish depuration plant, bacteriophage particles (model viruses) are not eliminated to any significant extent. Hence it can be expected that if filter-feeding shellfish have been exposed to contaminated waters, the risk of viral infection will persist even in mussels which are bacteriologically clean. The only safe procedure is therefore to advocate consumption of cooked shellfish, since this would inactivate the viral particles.

The work reported by the Authors on the study of distribution and survival of rotaviruses and enteroviruses in the marine environment is also very interesting, particularly in view of the fact that many of the cases of gastro-enteritis associated with swimming in polluted waters and consuming contaminated shellfish are known to be caused by viruses. In my opinion it is impractical, however, to try to apply a viral standard to either shellfish, their growing waters, or bathing waters. The techniques for viral enumeration are by no means standardized, nor simple enough for routine laboratories, and furthermore the standards themselves are not yet based on epidemiological evidence. A more sensible approach is to try to determine an indicator method for assessing the viral hazard; suitable candidates for this task might be sulphate-reducing clostridia, or perhaps F-specific bacteriophages.

DR LACK and MRS COOPER, Paper 14
With reference to Mr Paul's comments on calibration of the hydrodynamic and dispersion models, we recognize that field validation of the model is an important part of any programme. Preliminary trials of radar validation of the hydrodynamic model using the Marex OSCR technique suggest that, although the technique is more expensive than conventional drogue tracking, the results are much more

comprehensive for the surface water layers. The main disadvantage with radar at the moment is that we have not yet been able to establish how the surface water movements measured by radar relate to the depth-averaged currents predicted by the model. However, the technique is still in its early stages and the future is likely to see further interpretation of the results leading to our better understanding. Validation of the bacterial dispersion model is also of the utmost importance in any programme of outfall design. WRc are currently planning a piece of research evaluating and comparing four tracer techniques at the two case study outfalls described in the Paper.

Determination of scope for growth, rather than the direct measurement of growth itself, provides an immediate insight into the individual components which affect the changes in growth rate. High, positive scope for growth, coupled with increases in body weight, shows that the animal is not stressed to the extent that it is having to use energy reserves to meet the metabolic requirements for detoxication. Negative scope for growth suggests that the animal's metabolism has been stressed in dealing with contaminants and that the assimilative capacity of the ecosystem may be at its limits. In between these two extremes is a range of relative degrees of stress which must be compared with a known control site to interpret the environmental conditions which imposed that stress. The major advantage of the technique is that natural variations in conditions such as temperature and sea state are automatically integrated into the results, thus allowing the scientist to attribute signs of stress from natural or pollution-induced causes. Extensive laboratory tests are also in progress to further our knowledge of the direct effects of sewage, sludge and list 1 and 2 compounds.

Viruses are receiving much political scrutiny at present. Too little is currently known about the dispersion and survival of viruses in sea water, or about the health risks of bathing in sewage-contaminated sea water. Rotaviruses are major causative agents of gastro-intestinal illness in young children and are known, in some cases, to have spread water-borne disease. WRc have commissioned a programme of work with the Welsh Water Virology Unit to develop tehniques for detection of rotavirus in environmental samples and to incorporate that methodology into a study of the distribution and survival of viruses in sea water and sediments, with particular reference to bathing beaches.

To give quantitative expression to the health risk of bathing in sewage-contaminated waters, information is needed on the incidence of relevant diseases in a sample population and some means of linking these diseases with sea bathing. The last official UK study of this nature was reported by the Medical Research Council in 1959. It is probably appropriate for the UK to carry out an epidemiological study to assess the public health implications of the standards laid down in the EC bathing waters directive. The study should aim to answer two questions: what is the difference in incidence of disease between sea bathers and non-bathers, and what is the difference in incidence of diseases between sea bathers in clean and sewage-contaminated waters?

In reply to Mr MacDonald, we agree that the problem of storm water overflows is one which is important and frequently overlooked. WRc are currently developing a spill mode program for storm overflows to accompany the hydrodynamic and dispersion models described in the Paper. It is hoped that this will assist the engineer in the better design and location of storm water outfalls. The sea outfall should be regarded as an extension of the sewerage system and outfall design is best when it is part of an integrated sewerage design scheme which includes the storm overflows.

In reply to Mr McKemey, it is unfortunate that the results of ecological surveys must rely on educated opinion and are therefore difficult to explain to the non-scientist, especially the general public and politicians. The natural environment in which marine organisms live is often hostile and can be completely disrupted by storms or natural changes in tidal currents such as those experienced in an estuary over a spring-neap tidal cycle. Because of this, many benthic organisms tend to be opportunistic in their lifestyle, i.e. they can invade an area as soon as it becomes habitable and increase their numbers rapidly. These naturally large fluctuations in numbers make pollution-induced changes difficult to detect and virtually impossible to predict. Changes attributable to sewage outfalls are not always adverse in terms of total numbers of animals, although the distribution of species may change, resulting in a lower diversity index.

These difficulties have led to the recommendation that it is better to search for patterns among biological data and interpret them in terms of environmental data derived from surveys such as those recommended in the Paper. The methods used to reveal patterns and structure in data are multivariate statistics which aid, but do not substitute, informed interpretation. These methods are always, of necessity, retrospective and must be the nearest we can come to mathematical modelling.

In reply to Mrs Evison, although the epidemiology of diseases related to viruses in shellfish was not raised by us, we endorse the comments on viral standards and routine techniques for enumeration of viruses. The use of conservative tracers as indicators of viral contamination may well be suitable for sea water, but extrapolation to shellfish tissue assumes a relationship between viral numbers in water and shellfish tissue. Rates of uptake and depuration of bacteria by shellfish are dependent on water temperature and this would be reflected by the use of clostridia; F-specific bacteriophages may be more suitable indicators of viral contamination. In the case of bathing waters, work carried out by WRc suggests that the behaviour

of clostridia may be similar to that of viruses, particularly
with respect to their tendency to associate with sediment
particles. However, it may be over-cautious to assume that
viruses will also live indefinitely in sea water or
sediments.

DR CHARLTON and DR DAVIES, Paper 15

Mr Paul's first question concerned the application of
monitoring to the hydraulic behaviour of outfall diffuser
systems. As indicated in the Paper, our laboratory tests
showed that variations in the head loss characteristics of a
diffuser system can be used diagnostically to monitor its
performance. Field tests by ourselves and by WRc have shown
that these changes can be picked up on real operating
outfalls. These field tests also showed that many of the
head changes in a pumped outfall are considerably modified by
transient effects. However, where the operating conditions
of an outfall permit periods of steady state discharge there
is considerable scope for pressure head monitoring. In
practice the head measured will be the overall head loss
including the outfall pipe loss, as the only practical way to
monitor outfall head losses is to record the pressure head
difference between the outfall landward end, at a point where
the pipe or shaft is always running full, and the sea
surface. This will involve placing one pressure transducer
in a clean water well in the outfall pipe or shaft, and
another locally in the sea. The differential output from
these transducers is then recorded on the same time base as
the outfall discharge (which we hope would normally be
measured at some point in the system). The actual discharge
- head loss characteristic of the outfall - would be recorded
soon after the outfall was commissioned, but after the
initial sliming had occurred, and this characteristic then
used as a baseline for future monitoring operations.

The other question related to utilizing controlled
intrusion as a prediluting feature. The possibility of this
was demonstrated in the laboratory and used in a small two-
port outfall which had a considerable discharge variation and
would have intruded anyway (Fig. 20 in the Paper). In this
case the internal dilution was only about two times, due to
the low intrusion energy available from the short risers. In
outfalls with long risers, where the differential intrusion
head is considerable, there is rather more scope for
increasing this dilution ratio.

MR REYNOLDS and MR WILLIS, Paper 16

In response to the points raised by Mr Budd, the use of
sophisticated stability theories has to be weighed against
the advantages forthcoming from the results. In our
experience the theories shown in the Paper are adequate for
most cases and the use of more complex theories does not
present tangible cost savings in either materials or methods
of construction. Furthermore, it must be appreciated that in

most cases outfalls will be buried and, providing cover depths have been properly considered, should not be exposed to hydrodynamic forces during the required design life. In this context it may be more appropriate to concentrate on the stability and performance of backfill material.

The comments raised about security are particularly valid in areas where fishing takes place and in our view the best protection against trawl boards and spans is to bury the outfall or to provide protection by means of mattresses or stone.

Regarding stress analysis, allowable stress levels are a matter of choice and opinion which vary from specification to specification. In our opinion the values quoted in the Paper have proved to be satisfactory by experience and, although higher values can be considered, the cost effectiveness will not compensate for the reserves that are present in the more conservative values that are quoted.

With regards to Mr Budd's comments on construction techniques, the proof of the pudding where laybarges are concerned will be clearly shown by carrying out budget cost studies or during the competitive tendering stage in a project, whereby different construction methods can be compared. Savings that may be available due to mobilization costs for laybarges being shared between contracts will only become evident during the tender stage and cannot as a rule be budgeted for at an earlier time.

In response to Mr King, the possibility of total packages where the authority is not involved with capital expenditure should present financial advantages which are certainly worth evaluating. It is difficult, however, to generalize how the advantages will be felt and at this stage we are of the opinion that projects should be considered separately and on their own merits. The use of design and construct contracts is now well established and in most cases enables contractors to offer complete flexibility with regard to construction techniques and the associated commercial risks.

The use of ductile iron pipes for pulled outfalls compared with weldable steel has attractions where corrosion protection is concerned. However, the system is yet to be fully developed for long outfalls where the pulling loads would be high. No information is available regarding relative costs.

REFERENCES
1. Field J. G. et al. A practical strategy for analysing multispecies distribution patterns. Mar. Ecol., Prog. Ser. 8, 1982, 37-52.
2. Roddie B. Statistical analysis for marine survey data. WRc Report ER 1260M, 1986.
3. Cooper V. A. Environmental effects of discharges. Construction and maintenance of long sea outfalls, IPHE Technical and Training Symposium, Portsmouth, 1986, 6.1-6.11.

Closing address

D. O. LLOYD, BSc(Eng), FICE, Director, Sir William Halcrow & Partners

The first Paper of the Conference from Dr Mandl stated the
objectives of EEC policy as preserving, protecting and
improving the environment, as contributing to human health,
and as ensuring prudent use of natural resources. In
explaining the achievement of these objectives, he referred
to the difficulty in finding a common denominator in the EEC.
The future demands multilateral and bilateral action by
member states, but each country must remember that it is not
the only one and that it is not a question of bureaucratic
dictation.

Dr Davies from the US Environmental Protection Agency (EPA)
introduced some contentious viewpoints. The USA finds UK
attitudes complacent and Congress is busy adopting land-based
solutions, making it illegal to discharge sludge to sea by
pipe and to dump in shallow waters. Although 75% of the US
population will live within 30 miles of the coast in 10-20
years, could the EPA not be over-reacting, or not taking
sufficient account of the environmental impact of land-based
alternatives?

For the UK situation, Dr Matthews emphasized the need for
careful management of the sea, warning against misinformation
and scaremongering. Mr Wakefield's balanced approach
reflects the acceptable face of the anti-pollution lobby.

Current practice on outfalls in France, Australia and the
UK was covered. Australian attitudes to marine disposal
currently lie between those of the UK and the US, but seem to
be moving more to the US situation. Mr Huntington and Mr
Rumsey summarized the UK position with two points. Firstly,
the uses to which water receiving wastes is put should be
identified, then designed and operated accordingly, and
secondly, that if all discharges had to be treated in the UK,
capital costs per annum would double.

Of course, inland people want sea disposal, and coastal
people want land disposal.

Current practice on disposal by vessel was highlighted in
two Papers. Professor McIntyre addressed options and
management and identified three factors on sea disposal.
Material dumped as sea does not come back, there is no public
health risk, no effect on fishing, and no environmental

impact. Strategy and tactics, and effects were also
identified. The lack of proportion between the two dumping
options of dispersal and accumulation were also referred to.

Mr Harper introduced his mixed sludge strategy from North
West Water, and referred to the Liverpool Bay success story
and coined an intriguing new term, 'contamination
containment'. His co-Author Mr Greer promoted Strathclyde
Regional Council's summer sludge cruise and invited the EEC
and the US EPA to see for themselves that their disposal
treatments do work.

Four UK water authorities gave accounts of their work in
hand. All had common factors but some individual actions
stood out: Welsh Water dealing with problems on capital
expenditure constraints, Southern Water's detailed total
resource analyses, Wessex Water's struggle with risk factors,
and Thames Water's detailed and effective rebuttals of draft
EEC directives.

For the future, Mrs Cooper from WRc presented improved
methods now available for monitoring environmental impact.
Dr Charlton and Dr Davies commented on research and possible
important ways in which to prevent outfall problems. Mr
Reynolds and Mr Willis discussed investigation and design
(the unchanging items), and materials and construction
techniques (the developing items) for design and construction
of sea outfalls, including the turnkey option package.

Some of the messages from the Papers conflicted, but to
summarize overall, the following points are offered

(a) the best practical environmental solution must be
    found
(b) marine treatment of sewage and sludge is significantly
    more economic than are land-based options
(c) the public should be kept fully informed
(d) investigations should be done thoroughly
(e) the dangers of possible viral contamination problems
    and storm water should be heeded
(f) results should be monitored
(g) due notice of Continental views and the EEC should be
    taken.